BTEC FIRST AWARD

edexcel

D1434382

APPLICATION OF SCIENCE

David Goodfellow • Sue Hocking • Ismail Musa

PEARSON

Published by Pearson Education Limited, Edinburgh Gate, Harlow, Essex, CM20 2JE.

www.pearsonschoolsandfecolleges.co.uk

Copies of official specifications for all Edexcel qualifications may be found on the Edexcel website: www.edexcel.com

Text © Pearson Education Limited 2012
Edited by Ashwell Enterprises Limited and Priscilla Goldby
Designed by Wooden Ark Limited and Andy Magee
Typeset by Tech-Set Ltd, Gateshead
Original illustrations © Pearson Education Limited 2012
Illustrated by Tech-Set Ltd
Cover design by Pearson Education Limited
Picture research by Susie Prescott
Cover photo © Adam Gault/OJO Images/Getty Images

The rights of David Goodfellow, Sue Hocking and Ismail Musa to be identified as authors of this work have been asserted by them in accordance with the Copyright, Designs and Patents Act 1988.

First published 2012

16 15 14 13
10 9 8 7 6 5 4 3 2

British Library Cataloguing in Publication Data
A catalogue record for this book is available from the British Library

ISBN 978 1 446 90280 6

Printed in Italy by L.E.G.O. S.p.A.

A note from the publisher
In order to ensure that this resource offers high-quality support for the associated BTEC qualification, it has been through a review process by the awarding organisation to confirm that it fully covers the teaching and learning content of the specification or part of a specification at which it is aimed, and demonstrates an appropriate balance between the development of subject skills, knowledge and understanding, in addition to preparation for assessment.

While the publishers have made every attempt to ensure that advice on the qualification and its assessment is accurate, the official specification and associated assessment guidance materials are the only authoritative source of information and should always be referred to for definitive guidance.

BTEC examiners have not contributed to any sections in this resource relevant to examination papers for which they have responsibility.

No material from an endorsed student book will be used verbatim in any assessment set by BTEC.

Endorsement of a student book does not mean that the student book is required to achieve this BTEC qualification, nor does it mean that it is the only suitable material available to support the qualification, and any resource lists produced by the awarding organisation shall include this and other appropriate resources.

About this book

About your BTEC First Applied Science

Choosing to study for a BTEC First Applied Science qualification is a great decision to make for lots of reasons. More and more employers are looking for well-qualified people to work within the fields of science, technology, engineering and maths. The applied sciences offer a wide variety of careers, such as forensic scientist, drug researcher, medical physics technician, science technician and many more. Your BTEC will sharpen your skills for employment or further study.

Your BTEC First Applied Science is a vocational or work-related qualification. This doesn't mean that it will give you all the skills you need to do a job, but it does mean that you'll have the opportunity to gain specific knowledge, understanding and skills that are relevant if you go onto further study or into technician-related employment.

What will you be doing?

Application of Science is the second of the two BTEC Applied Science Awards. The first is Principles of Applied Science. In both Awards you will be covering all aspects of science, including biology, chemistry and physics, as well as maths and health and safety-related issues. You will complete assignments based on scientific job-related scenarios, for example, working as a science technician, producing information for the public and conducting scientific research. As well as exploring a range of scientific concepts, you will use your IT skills to produce documents for assessment. Other skills you will practise include researching, preparing and giving presentations, producing scientific reports, following instructions for practical investigations and effective time management.

About the authors

David Goodfellow is a freelance writer and examiner, having previously taught chemistry at all levels for over 20 years. He led the development of the AS Science qualification for 2007 and is an experienced author.

Sue Hocking has been an examiner for almost 30 years. She has delivered BTEC science and health studies courses in FE colleges, as well as GCSE and A level biology courses in secondary schools and a sixth form college. Her specialist fields are biology, biochemistry and health promotion. Sue was the series editor for OCR A level Biology and has written many books and teacher support resources.

Dr Ismail Musa is a standards verifier for BTEC Applied Science (Level 2 and 3), an examiner for both GCSE and A level Physics and was involved in the development of the 2010 specification for BTEC Applied Science. Ismail has been teaching vocational applied science courses for over 12 years. As well as teaching and examining he has been working as a Subject Learning Coach for Science (SLC), coaching students and staff and organising and delivering teaching and learning sessions.

Contents

How to use this book

This book is designed to help you through your BTEC First Applied Science Award in Application of Science. It is divided into four units to reflect the units in the specification. This book contains many features that will help you use your skills and knowledge in work-related situations and assist you in getting the most from your course.

Introduction

These introductions give you a snapshot of what to expect from each unit – and what you should be aiming for by the time you finish it.

Learning aims

Learning aims describe what you will be doing in this unit.

Learner voice

A learner shares how working through the unit has helped them.

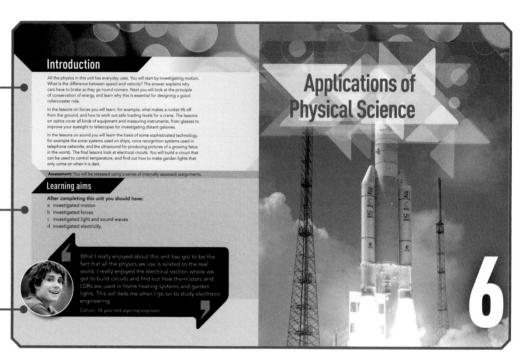

Get started

Get started with a short activity or discussion about the lesson.

Key terms

Key terms boxes give definitions of important words and phrases that you will come across.

Worked example

Worked examples provide a clear idea of what is required for a calculation.

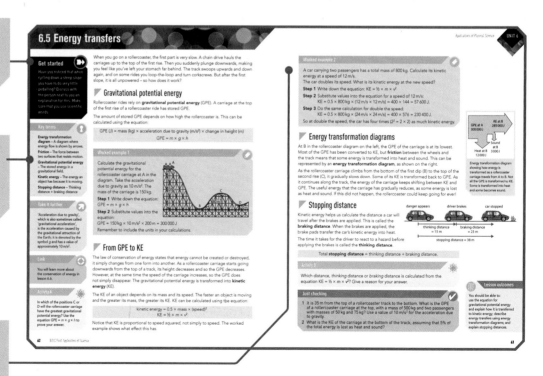

You will be assessed in two different ways for your BTEC First Applied Science Award. For Units 5–7 your teacher will set assignments for you to complete. The table in the BTEC Assessment Zone for Units 5–7 explains what you must do in order to achieve each of the assessment criteria.

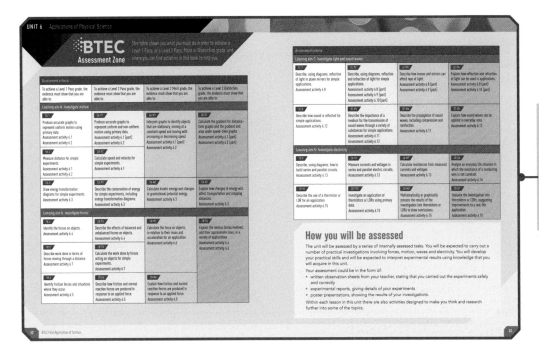

Assessment criteria

This table in the BTEC Assessment Zone signposts assessment activities you'll find in this book to help you to prepare for your assignments.

For Unit 8 you will be assessed externally using a paper-based test. The Assessment Zone helps you to prepare for the test by showing you some of the different types of questions you will need to answer.

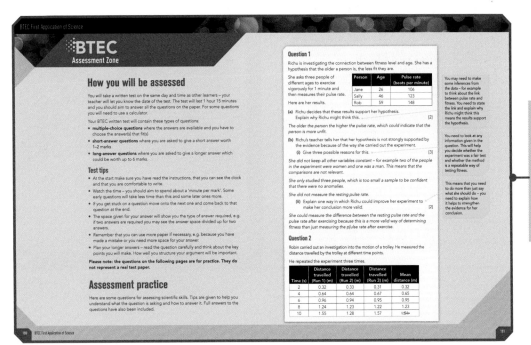

Assessment Zone

The Assessment Zone at the end of Unit 8 contains information on how you will be assessed. It also includes practice questions to help you prepare for your external test.

There are different types of activities for you to do to help you develop your knowledge, skills and understanding.

Activities

Activities will help you show the knowledge and understanding you have gained in the lesson.

Assessment activities

Assessment activities are suggestions for tasks that you might do to help build towards your assignment. Each Assessment activity has **Tips** which provide guidance on how to achieve the best grade.

Just checking

Use these to check your knowledge and understanding at the end of the lesson.

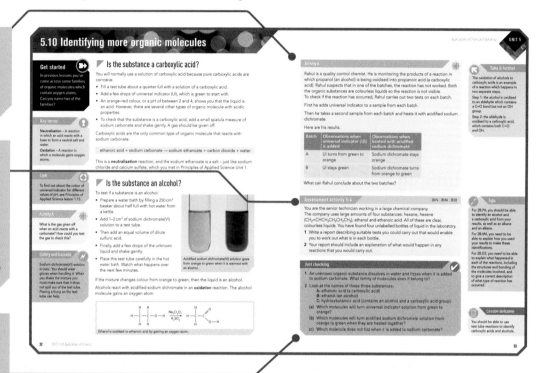

Case studies

Case studies show you examples of specific scientific topics in the workplace.

WorkSpace

WorkSpaces provide a snapshot of someone who works in the industry and of real workplace issues. They show how the skills and knowledge you develop can help you in your career.

Think about it

WorkSpaces also give you the chance to think more about the role that this person does, and whether you want to follow in their footsteps once you've completed your BTEC.

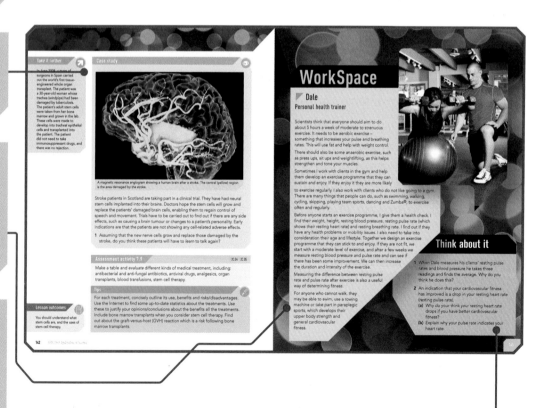

Planning and getting organised

The first step in managing your time is to plan ahead and be well organised. You can improve your planning and organisational skills by:

- Using a diary to schedule all the work you have to do. You could use this as a 'to do' list and tick off each task as you go.
- Dividing up long or complex tasks into manageable chunks and putting each 'chunk' in your diary with a deadline of its own.
- Always allowing more time than you think you will need for a task.

Organising and selecting information

Once you have gathered a range of information during your research, you will need to organise the information so it's easy to use.

- Make sure your written notes are neat and have a clear heading – it's often useful to date them, too.
- Always keep a note of where the information came from (the title of a book, the title and date of a newspaper or magazine or the web address of a website) and, if relevant, which pages.
- Work out the results of any questionnaires you've used.

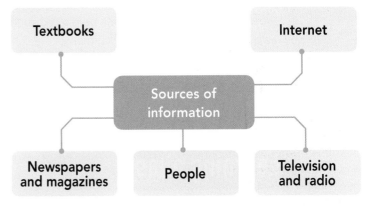

Once you have completed your research, re-read the assignment brief or instructions you were given to remind yourself of the exact wording of the question(s) and divide your information into three groups:

1 Information that is totally relevant.
2 Information that is not as good, but which could come in useful.
3 Information that doesn't match the questions or assignment brief very much, but that you kept because you couldn't find anything better!

Check against the questions or assignment brief that there are no obvious gaps in your information. If there are, make a note of them so that you know exactly what you still have to find.

Presenting your work

Before handing in any assignments, make sure:

- you have addressed each part of the question and that your work is as complete as possible
- all spelling and grammar is correct
- you have referenced all sources of information you used for your research
- that all work is your own – otherwise you could be committing **plagiarism**
- you have saved a copy of your work.

Introduction

Our society today depends so much on the fossil fuels obtained from crude oil. In this unit, you will find out more about how these fuels produce energy when they burn, as well as how crude oil is processed to make these fuels.

You will carry out investigations to measure the energy produced from burning different fuels and also look at some other reactions which produce energy changes.

Crude oil is also what the chemical industry uses to make an incredible range of other useful products. You will study some of these compounds in the laboratory and also find out why they are so important for society.

Some of the newest materials being produced are nanochemicals and you will find out more about this exciting kind of chemistry.

Chemists must always be prepared to assess the risks of any new substance, and you will be doing this as well as evaluating whether the use of some of these substances can be justified.

Assessment: You will be assessed using a series of internally assessed assignments.

Learning aims

After completing this unit you should have:

a investigated and understood enthalpy changes associated with chemical reactions

b investigated organic compounds used in society

c explored the uses of nanochemicals and new materials.

I knew that we get lots of fuels from crude oil but I didn't realise how many other things come from crude oil as well. I was really interested to find out that nanoparticles are in so many things that people use every day, like cosmetics. It's really important that people should understand the risks and benefits of using chemicals like this.

Nina, *15 years old*

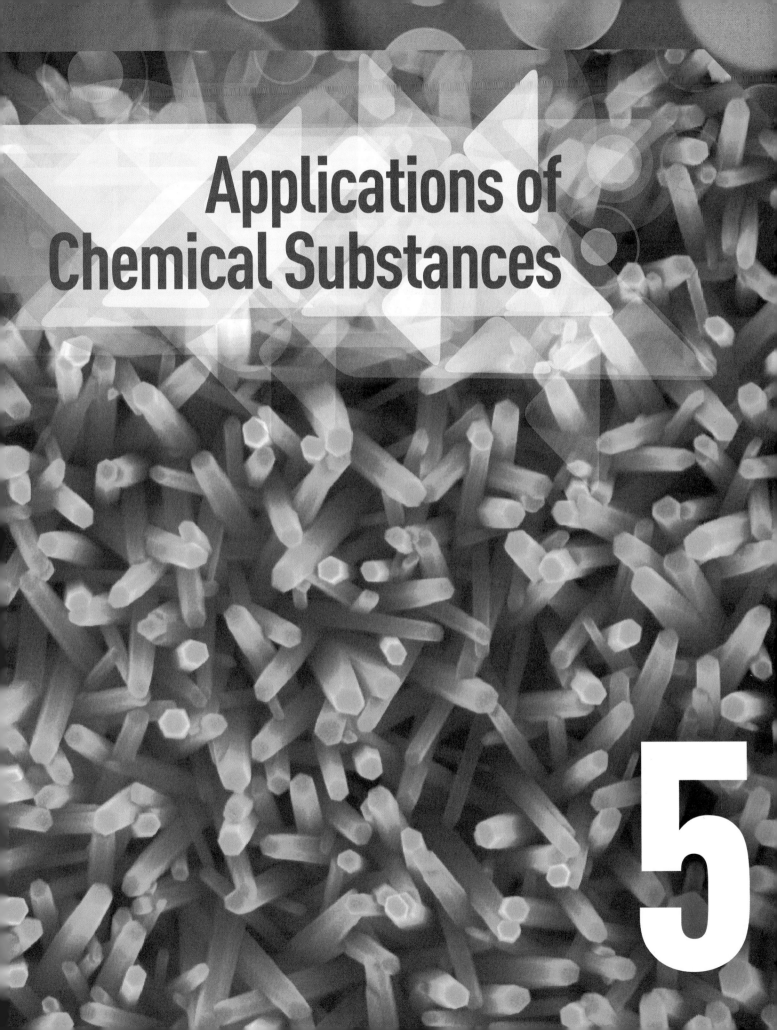

Applications of Chemical Substances

5

BTEC
Assessment Zone

This table shows you what you must do to achieve a Level 1 Pass, or a Level 2 Pass, Merit or Distinction grade, and where you can find activities in this book to help you.

Assessment criteria			
To achieve a Level 1 Pass grade, the evidence must show that you are able to:	To achieve a Level 2 Pass grade, the evidence must show that you are able to:	To achieve a Level 2 Merit grade, the evidence must show that you are able to:	To achieve a Level 2 Distinction grade, the evidence must show that you are able to:
Learning aim A: Investigate and understand enthalpy changes associated with chemical reactions			
1A.1 Measure the temperature changes associated with chemical reactions. Assessment activity 5.1	**2A.P1** Investigate temperature changes associated with exothermic and endothermic reactions using primary data. Assessment activity 5.1	**2A.M1** Explain why an overall reaction is exothermic or endothermic. Assessment activity 5.1	**2A.D1** Calculate the energy changes that take place during exothermic and endothermic reactions. Assessment activity 5.1
Learning aim B: Investigate organic compounds used in society			
1B.2 Identify the uses of the main fractions from the distillation of crude oil. Assessment activity 5.2	**2B.P2** Describe the fractional distillation of crude oil to produce a range of useful products. Assessment activity 5.2	**2B.M2** Explain how fractional distillation separates compounds due to different boiling ranges. Assessment activity 5.2	**2B.D2** Analyse the relationship between the boiling range and the length of carbon chain of fractions. Assessment activity 5.2
1B.3 Name alkanes and alkenes from structural and displayed formulae. Assessment activity 5.3	**2B.P3** Draw accurately the structural and displayed formulae of organic molecules. Assessment activity 5.3 Assessment activity 5.5 (part)	**2B.M3** Describe the bonding and structure of organic molecules. Assessment activity 5.3	**2B.D3** Explain the results of experiments to identify organic compounds in terms of their reaction type, structural and displayed formulae, and bonding. Assessment activity 5.4
1B.4 Identify an alkene and an alkane using primary observations. Assessment activity 5.4	**2B.P4** Identify an alkene and a carboxylic acid using primary observations. Assessment activity 5.4	**2B.M4** Explain how a series of experiments can be used to identify organic compounds based on their solubility and reactions. Assessment activity 5.4	
1B.5 Identify uses of ethene, ethanol and ethanoic acid. Assessment activity 5.5	**2B.P5** Describe the uses of organic compounds in our society. Assessment activity 5.5	**2B.M5** Explain the problems associated with the use of organic molecules. Assessment activity 5.5	**2B.D4** Evaluate the benefits and drawbacks of using organic materials. Assessment activity 5.5

Learning aim C: Explore the uses of nanochemicals and new materials

1C.6	2C.P6	2C.M6	2C.D5
Define nanochemicals. Assessment activity 5.6	Describe a use of nanochemicals, smart and specialised materials. Assessment activity 5.6	Explain the benefits of using nanochemicals, smart and specialised materials. Assessment activity 5.6	Evaluate the benefits and drawbacks of using nanochemicals, smart and specialised materials. Assessment activity 5.6

How you will be assessed

The unit will be assessed by a series of internally assessed tasks. You will be expected to show an understanding of chemistry relevant to industrial processes, environmental issues and the use of chemicals across a wide range of industries.

The tasks will be based on various scenarios that place you in the position of working in the industrial sector – for example, in the research and development department of a chemical firm, as a laboratory technician or in an oil refinery.

Your actual assessment could be in the form of:

- written materials such as a training manual or report
- a log of experimental observations and results
- a presentation or DVD describing important scientific principles.

5.1 Energy changes in chemical reactions

Get started

When something burns, what types of energy are given out? Where has the energy come from?

Key terms

Endothermic – A reaction that takes in energy (in the form of heat) from the surroundings, causing a fall in temperature.

Exothermic – A reaction that gives out energy (in the form of heat) to the surroundings, causing a rise in temperature.

The thermite reaction produces enough heat to melt the iron that is produced.

Exothermic and endothermic reactions

Reactions that *give out* energy are called **exothermic** reactions. In exothermic reactions, the temperature increases.

Reactions that *take in* energy are called **endothermic** reactions. In endothermic reactions, the temperature falls.

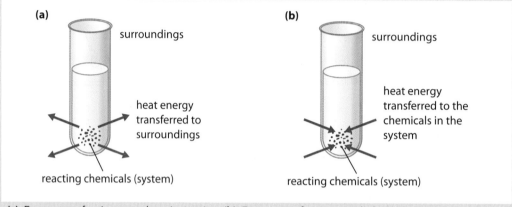

(a) Energy transfers in an exothermic reaction. (b) Energy transfers in an endothermic reaction.

The **thermite reaction** is a dramatic example of an exothermic reaction. In this reaction, aluminium reacts with iron oxide to produce iron and aluminium oxide. Energy, in the form of heat and light, is given out and the temperature of the surroundings increases dramatically. So much heat is given out that it melts the iron that is formed.

The thermite reaction is used as a portable source of molten iron to weld together pieces of steel, like damaged railway tracks. It has even been used for welding underwater, as the reaction doesn't need air.

Cold packs, used to treat some injuries, contain chemicals which react together in an endothermic reaction.

Worked example

James is a chemical technician working for an electronics company. He is investigating how to produce hydrogen that can be used to power fuel cells. The fuel cells are for use in an emergency in remote locations where mains electricity is not available.

James adds a piece of magnesium ribbon to some hydrochloric acid solution. The temperature of the acid solution goes up from 22 °C to 45 °C.

What can James conclude from this information?

Step 1 He can calculate the temperature change: 45 − 22 = 23 °C.

Step 2 The temperature went up, so the temperature change is positive: +23 °C.

Step 3 A positive temperature change means that heat is given out by the reaction, so the reaction is exothermic.

Dissolving reactions

What happens when you **dissolve** different substances in water?

When sodium carbonate is added to water, an exothermic reaction happens. Heat is given out by the reaction and, because the water is in contact with the reaction, this heat is transferred to the water. So the temperature of the water goes up, which is a positive temperature change.

Dissolving ammonium chloride in water is an endothermic reaction. This time heat is absorbed (taken in) by the reaction. The water is in contact with the reaction so this heat is transferred from the water to the ammonium chloride. So the temperature of the water will go down, which is a negative temperature change.

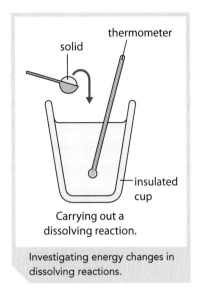

Carrying out a dissolving reaction.

Investigating energy changes in dissolving reactions.

You might think that if heat is absorbed during a dissolving reaction, then the ammonium chloride should get hotter! But this won't happen, because in an endothermic reaction the heat doesn't stay as heat – it is transferred into stored energy in the chemical bonds.

You can see in the diagram on page 14 the energy transfers which happen between reacting chemicals and the water which is in contact with them.

Activity A

Ajay is going to make a poster about exothermic and endothermic reactions to put on the wall of his student chemistry laboratory. He carries out some experiments. He dissolves some ionic solids in water and investigates the temperature changes.

Substance	Starting temperature (°C)	Highest or lowest temperature reached (°C)
Calcium chloride	21	35
Ammonium nitrate	19	1
Sodium chloride	21	20

1 Calculate the temperature change in each experiment and say whether it is positive or negative.

2 Use the results to decide whether each reaction is exothermic or endothermic.

Just checking

1 What is the difference between an exothermic and an endothermic reaction?

2 Give one example of an exothermic reaction and one example of an endothermic reaction.

Lesson outcomes

You should be able to recognise exothermic and endothermic reactions and how to classify temperature changes during reactions as positive or negative.

5.2 Explaining energy changes

The energy of the chemicals in a reaction is called the enthalpy. The energy change is sometimes called the **enthalpy change**.

The overall enthalpy change for a reaction depends on the energy changes which happen when bonds are broken and formed.

Breaking bonds and making bonds

When chemical reactions happen, there is a change in the way in which the particles are bonded. Most reactions involve some bonds being broken and some bonds being made.

- Breaking bonds requires energy.
- Making bonds releases energy.

This is the reaction that takes place when methane burns in air:

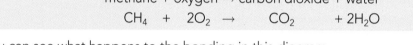

$$\text{methane} + \text{oxygen} \rightarrow \text{carbon dioxide} + \text{water}$$
$$CH_4 + 2O_2 \rightarrow CO_2 + 2H_2O$$

You can see what happens to the bonding in this diagram.

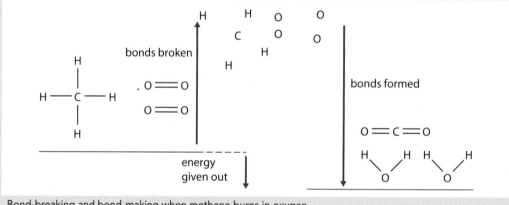

Bond-breaking and bond-making when methane burns in oxygen.

You can use ideas about bond-breaking and bond-making to explain why the reaction between methane and oxygen is exothermic.

- Some bonds are broken and others are made.
- Bond-breaking needs energy and bond-making gives out energy.
- Quite a lot of energy is needed to break the bonds in the methane and oxygen molecules.
- But even more energy is given out when new bonds form to make water and carbon dioxide molecules.
- So, overall the reaction gives out heat – the total of all the energy changes is exothermic.

Dinitrogen tetroxide (N_2O_4) is a pale yellow gas. It is mixed with other compounds to make a rocket propellant.

When N_2O_4 is heated, it **decomposes** to form nitrogen dioxide (NO_2). This reaction is reversible:

$$N_2O_4 \rightleftharpoons 2NO_2$$

The forward reaction is endothermic. During the reaction, a covalent bond is broken. You can see this if you look at the structures of the reactants and products.

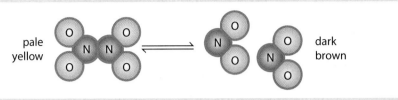

The covalent bond between the two nitrogen atoms is broken when N_2O_4 decomposes.

- Breaking bonds requires energy.

That is why the forward reaction is endothermic – the reaction takes in heat which is used to break the bond.

Now think about the reverse reaction, where two NO_2 molecules join together to form N_2O_4. In this process a covalent bond is made.

- Making bonds releases energy.

So the reverse reaction is exothermic.

Case study

Darren is a researcher for a small research group developing new rocket propellants for companies who want to launch small communications satellites.

'The main thing that companies are looking for is performance – the satellite payload is very expensive, so the fuel itself must be as reliable as possible. We are looking for fuels which react to produce a lot of heat and a lot of gas very quickly. That is what produces the thrust to lift the rocket. There's no oxygen in space so we also have to carry an oxidiser to supply the oxygen which the fuel needs to burn.'

Dinitrogen tetroxide (N_2O_4) can be mixed with hydrazine (N_2H_4) to make a liquid rocket fuel. When the fuel mixture is ignited, it reacts to produce a large volume of hot gas.

1 Why is it an advantage to use liquids as rocket fuels rather than gases?

2 Use the information above to decide whether the reaction between N_2O_4 and N_2H_4 is exothermic or endothermic.

3 Suggest some of the gases which might be formed when these substances react together.

The energy released by the reaction of the fuel in this rocket provides the energy for it to lift off.

Just checking

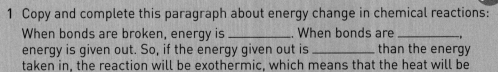

1 Copy and complete this paragraph about energy change in chemical reactions:
When bonds are broken, energy is _____. When bonds are _____, energy is given out. So, if the energy given out is _____ than the energy taken in, the reaction will be exothermic, which means that the heat will be _____.

2 When iron chloride dissolves in water, the temperature of the solution increases. Use ideas about bond-breaking and bond-making to explain why heat is released to the water when the iron chloride dissolves.

Lesson outcomes

You should be able to explain heat and enthalpy changes using ideas about bond-breaking and bond-making, and understand that the overall enthalpy change is a combination of the bond-breaking and bond-making enthalpy changes.

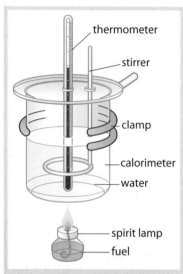

Using a fuel to heat water in a calorimeter allows you to calculate the energy released.

labels: thermometer, stirrer, clamp, calorimeter, water, spirit lamp, fuel

Burning fuels

A simple way to measure the energy change when fuel is burnt is to burn the fuel in a spirit burner and use this to heat some water in a container called a calorimeter.

In calorimetry experiments, heat is transferred to the water in the **calorimeter**. Scientists and technicians use the following equation to calculate how much energy is transferred to the water:

$$q = m \times c \times \Delta T$$

where

q = energy transferred, measured in **joules** (J)
m = mass of water (g)
c = **specific heat capacity** of water = 4.2 J/K/g
ΔT = temperature change (K).

Although the equation involves the mass of water, this doesn't mean you have to weigh the water. Because the density of water is 1 g per cm³, 200 cm³ of water has a mass of 200 g.

Different substances have different values of specific heat capacity. When given in units of J/K/g, this value tells you how much energy is needed to heat up 1 g of the substance by 1 K.

You might not recognise the unit 'K'. K stands for **Kelvin** and is the scientific unit of temperature. A change of 10 K is the same as a change of 10 °C. This means you can measure temperature changes using an ordinary Celsius thermometer.

Worked example

Sunita is a chemical technician in a laboratory investigating combined heat and power systems. She measures the energy released by different fuels. Here are some results taken from Sunita's investigations where she was using ethanol as a fuel.

Mass of ethanol burned = 1.3 g
Mass of water in calorimeter = 200 g
Temperature of water at start of experiment = 21 °C
Temperature of water at end of experiment = 44 °C

The temperature rise of the water in Sunita's experiment is 44 − 21 = 23 °C. Calculate the energy released by the fuel in Sunita's experiment.

Step 1 Use the equation: $q = m \times c \times \Delta T$
Step 2 Substitute the values above into the equation
Energy transferred = 200 g × 4.2 J/K/g × 23 °C = 19 320 J
Step 3 Divide by 1000 to change to kJ (kilojoules) = 19.32 kJ

To compare the ethanol fuel with other fuels, you can calculate the energy output in kilojoules per gram:

Energy output = energy transferred divided by mass burnt
= 19.32 kJ/1.3 g
= 14.86 kJ/g

Neutralisation reactions

Abby is a chemical engineer designing a pilot plant for a **neutralisation reaction**. She needs to calculate the energy given out when the reaction happens so that she can work out how to cool the plant safely. She puts 50 cm³ of sodium hydroxide in a polystyrene cup and measures its temperature. She adds 50 cm³ of hydrochloric acid and stirs. The temperature starts to go up and Abby measures the highest temperature reached.

Temperature of solution before the reaction = 19°C
Highest temperature reached during the reaction = 52°C
Temperature change = +33°C

Measuring the temperature change of reacting solutions.

This is a positive temperature change – the energy released by the reaction has been transferred to the solution.

Activity A

Use the same method as in the worked example to calculate the energy transferred in Abby's reaction.

Remember

Make sure that you use the total volume of water in the solutions which are reacting. Abby uses 50 cm³ of each solution, so the total volume is 100 cm³. 100 cm³ of solution contains 100 cm³ of water.

Most neutralisation reactions are exothermic, but not all. The reaction between citric acid and sodium hydrogen carbonate is endothermic and also results in a lot of carbon dioxide gas being produced. The combination of the fizzing and the cooling effect of the reaction provides a pleasant sensation when this combination of chemicals is used in confectionary like sherbet powder.

The neutralisation reaction which occurs between the chemicals in sherbet makes your mouth feel cool as it is endothermic.

Activity B

When 5 g of sodium carbonate was added to 100 cm³ of water, the temperature of the water went up from 21 °C to 24 °C.

1 Calculate the heat absorbed by the water.
2 Calculate the 'energy output' of sodium carbonate in kilojoules per gram.

Just checking

When 5 g of iron chloride was added to 200 cm³ of water in a calorimeter, it dissolved and the temperature changed from 18 °C to 39 °C.
1 Work out the temperature change. Is this an exothermic or an endothermic reaction?
2 Calculate the heat energy released in the experiment.

Lesson outcome

You should know how to use the equation $q = m \times c \times \Delta T$ to calculate the heat absorbed by water during reactions.

5.4 Hot and cold packs and energy profiles

Get started

Have you ever had an injury that was treated by applying heat or cold? What type of injury was it? How do you think the treatment helped?

Key term

Energy profile diagram – A way of displaying the changes in enthalpy (energy) which happen during a chemical reaction.

Activity A

Use information from lessons 5.1 and 5.2 to suggest other combinations of substances that could be used in:

1 a heat pack
2 an instant cold pack.

Applying an ice pack to an injured ankle may reduce swelling.

Heat packs

If you have a sports injury or other muscle pain, you can relax the muscles and loosen the tissues with a heat pack before taking more exercise. This also increases the blood supply to the injured area. Some heat packs use exothermic reactions to generate the heat.

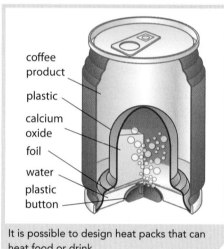

coffee product
plastic
calcium oxide
foil
water
plastic button

It is possible to design heat packs that can heat food or drink.

Single-use cool pack

INSTANT COLD PACK

ammonium nitrate
water

BREAK 'N' SHAKE TO USE
SINGLE-USE ONLY

When the user presses on the pad, the compartment holding ammonium nitrate breaks open, and the chemicals mix and react.

Instant cold packs

Some injuries, such as bruising or a sprain, can be treated by applying a cold pack to the injured area immediately after the injury occurs. This is to reduce swelling. Instant cold packs work in a similar way to heat packs, but solids like ammonium nitrate are used. These dissolve in water in an endothermic reaction.

Unlike ice-based cold packs, instant cold packs do not need to be kept in a freezer before use.

Energy profile diagrams

Energy profile diagrams show what happens to the bond energy or enthalpy of chemical substances during reactions.

The reaction between hydrochloric acid and sodium hydroxide to produce sodium chloride and water is an exothermic reaction. This means that less energy is absorbed breaking the chemical bonds in the **reactants** than is given out making the chemical bonds in the **products**.

So there is an overall enthalpy loss by the chemicals in an exothermic reaction. The energy is transferred as heat to the water in contact with the reaction.

The reaction between ammonium chloride and water is an endothermic reaction. This means that more heat is absorbed breaking the chemical bonds in the reactants than is given out making the chemical bonds in the products.

So there is an overall enthalpy gain by the chemicals in an endothermic reaction. The heat in the water is transferred as chemical energy in the bonds of the chemicals.

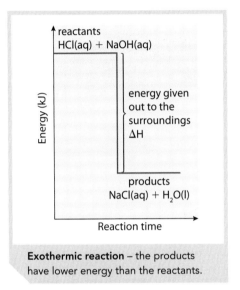

Exothermic reaction – the products have lower energy than the reactants.

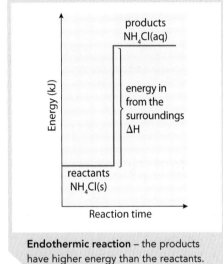

Endothermic reaction – the products have higher energy than the reactants.

Activity B

Draw an energy profile diagram for the combustion of methane in air. Use the information in lesson 5.2 to help you.

Assessment activity 5.1

| 2A.P1 | 2A.M1 | 2A.D1

You are a chemical technician working for a company that is intending to start making heat packs and cold packs, based on the dissolving of salts. You have been asked to find out which substances to put into these new products.

1 Plan investigations to identify suitable substances. Remember to check your plans with your teacher before you carry out any investigations.

2 Prepare a presentation to communicate your findings.

Tips

Make sure that you measure the temperature changes in both an exothermic and an endothermic reaction.

For 2A.P1, you will need to measure the temperature changes for a number of different reactions and use your results to explain the meaning of the terms 'exothermic' and 'endothermic'.

For 2A.M1, you will need to explain the energy changes during these reactions, using ideas about heat absorbed or lost by the water and the chemicals.

For 2A.D1, you will need to calculate the actual energy absorbed or supplied by the water – this will help you to decide on the best substances to use in your packs. You will also need to use ideas about bond-breaking and bond-making to explain the energy changes in these reactions.

Just checking

1 Explain why endothermic reactions are used in cold packs.

2 The dissolving of calcium chloride is used in a heat pack because it is an exothermic reaction. Look at the following statements about this reaction and decide whether they are true or false.
   A The energy released by forming bonds is greater than the energy needed to break bonds.
   B The enthalpy of the chemicals increases during the reaction.
   C Energy is transferred from the water to the bonds in the chemicals.

Lesson outcomes

You should understand how heat packs and cold packs work, and you should be able to use energy profile diagrams to show the energy (enthalpy) change in reactions.

Get started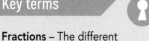

Crude oil was formed from the remains of living creatures which fell to the bottom of the sea millions of years ago. What happened to turn these remains into crude oil? Why is crude oil always found underground?

Key terms

Fractions – The different component parts of crude oil.

Hydrocarbons – Chemicals that contain only hydrogen and carbon atoms.

Crude oil is a vital resource for the modern world. It is a mixture of many different substances. Some of these substances are used as fuels in industry and transport. Others are used as the starting materials in important reactions to produce products like plastics, detergents and paints.

Fractional distillation

Because crude oil is a mixture, it can be separated into simpler substances, called **fractions**, according to their boiling points. The method used to do this is called **fractional distillation**.

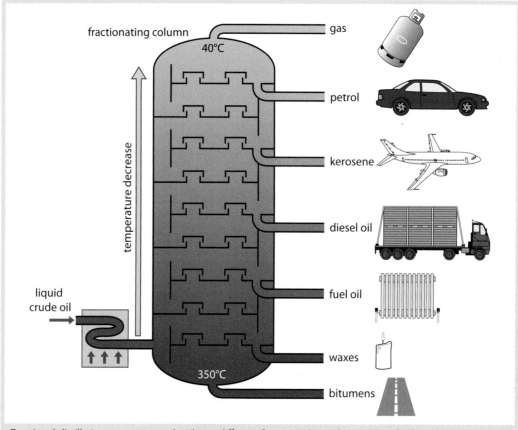

Fractional distillation separates crude oil into different fractions depending on their boiling points.

- The crude oil is heated and becomes a vapour.
- The vapour passes into a fractionating column.
- The vapour rises up the column and cools down.
- The substances in the vapour start to condense into liquids. Substances with a high boiling point condense first.
- Some substances have such a low boiling point that they never condense into a liquid and pass out of the column as gases.

The fractional distillation column shown above is designed to separate crude oil into seven different fractions.

Each fraction is not actually a pure substance. It will contain a small number of different molecules which boil at similar temperatures.

Activity A

1 Use the Internet to find animations of fractional distillation. These will help you understand how a fractionating column works.

2 Put these fractions in order of their boiling point, starting with the lowest: bitumens, diesel oil, gas, kerosene.

Hydrocarbons as fuels

The molecules in crude oil are mainly **hydrocarbons** – they contain hydrogen and carbon atoms only. Some of these hydrocarbons contain very long chains of carbon atoms while others are quite small molecules.

Hydrocarbons will burn in air to produce heat. Carbon dioxide and water are also formed.

The simplest hydrocarbon is methane. It is a gas and is easily transported through pipes to be used for heating and cooking in homes and factories.

Methane burns in air:

$$CH_4 + 2O_2 \rightarrow CO_2 + 2H_2O$$
methane + oxygen → carbon dioxide + water

Propane and butane are also gases. They are put under pressure which converts them into liquids, and are sold as bottled gas for camping stoves.

A liquefied mixture of propane and butane is also known as LPG (liquefied petroleum gas). It is used instead of petrol in cars that have specially converted engines.

All the other hydrocarbon fuels (such as petrol, kerosene and diesel) are liquids. You can find out more about these in the next lesson.

This stove uses a mixture of propane and butane as a fuel.

Safety and hazards

When hydrocarbons are used as fuels, it is important to make sure that there is a good supply of air. If there is not enough oxygen, the reaction will produce carbon monoxide as well as carbon dioxide. Carbon monoxide is a toxic gas and is fatal if it builds up in the body. It bonds permanently to the haemoglobin in red blood cells so oxygen cannot be transported round the body. Carbon monoxide is colourless and odourless so it is difficult to know when it is present. Carbon monoxide detectors can be used to help keep you safe.

Carbon monoxide detector.

Just checking

1 Look at the following statements about fractional distillation. They all contain a mistake. Write out the statements again so that they are correct.

   A Crude oil is heated to turn it into a liquid.

   B As the vapour rises up the fractionating column it gets hotter and starts to condense.

   C The substances with the highest boiling points condense at the top.

   D Substances with very low boiling points exit the top of the column as liquids.

Lesson outcome

You should know how fractional distillation is used to separate crude oil into different fractions, some of which are used to produce fuels.

Get started

Crude oil contains hydrocarbons. Where did the carbon and hydrogen atoms in these molecules come from?

Structure of hydrocarbons

When crude oil is distilled in a fractionating column, the different fractions contain hydrocarbons with different numbers of carbon atoms in their chains.

(a) propane: C_3H_8

(b) dodecane: $C_{12}H_{26}$

(a) Propane has three carbon atoms in a chain and is found in the gas fraction.
(b) Dodecane has 12 carbon atoms and is found in the kerosene fraction.

Molecules with longer hydrocarbon chains have higher boiling points. There are stronger forces between the molecules, so more energy is needed to separate them.

They are also more **viscous** (thick and treacly) and burn less easily.

Uses of fractions

The different fractions have different uses, as shown in the table.

Fraction	Boiling range	Number of carbon atoms in the chain	Use(s)
Gas	Below 40 °C	1–4	Domestic fuel – 'natural' gas, i.e. methane Bottled gas – propane and butane
Petrol	40–200 °C	4–11	Fuel for car engines Raw materials for the chemical industry
Kerosene	180–250 °C	11–16	Fuel for aircraft
Diesel oil	200–360 °C	12–20	Fuel for diesel engines
Fuel oil	350–400 °C	20–27	Fuel for power stations Fuel for ships
Waxes	Above 400 °C	25–40	Candles Lubricants
Bitumen	Above 400 °C	More than 30, sometimes arranged in rings	Road building

The engine of this ship burns fuel oil.

Did you know?

Some people are now converting the petrol engines in their cars to run on LPG – liquefied petroleum gas. LPG contains hydrocarbon molecules, with three or four carbon atoms in the chain, which are stored under pressure to keep them liquid. In 2012, a litre of LPG cost less than half as much as a litre of petrol. LPG fuel also burns more cleanly, producing fewer particulates and lower carbon monoxide emissions.

Crude oil from different sources may contain different proportions of these fractions.

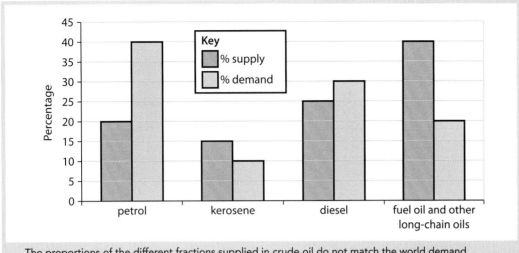

The proportions of the different fractions supplied in crude oil do not match the world demand for these fractions.

Activity A

Eicosane is a hydrocarbon with 20 carbon atoms in its chain.
1 Name two fractions that might contain eicosane.
2 Give the uses of these fractions.
3 Predict the likely boiling point of eicosane.
4 Look up the boiling point of eicosane in a databook or on the Internet. Were you right?

Case study

Rhys is a chemical engineer working for a petrochemical company. His job is to increase the supply of fractions that are in high demand. He runs a plant that carries out a reaction called cracking. This breaks down large molecules into smaller ones.

1 Identify one fraction for which demand outstrips supply.
2 Give one reason why there is a high demand for this fraction.
3 Suggest how Rhys will increase the supply of this fraction.

Tips

You should base your handout on a large diagram of a fractional distillation column.

For 2B.P2 you will need to give the uses of each fraction, as well as adding some labels to your diagram to explain how fractional distillation works.

For 2B.M2, you will need to add information about the boiling ranges of the fractions to help you in your explanation.

For 2B.D2, you should add information about the length of carbon chains, and use this to describe and explain the link between boiling point and length of chain.

Assessment activity 5.2

| 2B.P2 | 2B.M2 | 2B.D2

You are an education officer working for an oil refinery. You will be doing a talk for some school students who are visiting the refinery. Prepare a handout that you can give to the students to accompany your talk. The handout should describe the process of fractional distillation and explain how the fractions will be used.

Lesson outcomes

You should know the main uses of the fractions of crude oil and understand the link between the boiling point of hydrocarbons and the length of the hydrocarbon chain.

Just checking

1 Give the main use of these fractions: kerosene, fuel oil, bitumen.
2 What happens to the boiling point of a hydrocarbon as the chain length increases?

5.7 Organic molecules: hydrocarbons

The **organic molecules** in crude oil are hydrocarbons. There are several different families of hydrocarbon molecules.

Alkanes

Alkanes contain long chains of carbon atoms bonded together by single covalent bonds.

Here are the first four members of the alkane family.

All of the bonds between carbon atoms in alkanes are single bonds.

These diagrams are called **displayed formulae** – they show all of the atoms and all of the bonds so it is easy to see the structure of the molecule. Each of the lines between the atoms in the formula is a single bond. The two atoms are sharing a pair of electrons.

There are other ways of showing the formulae of organic molecules. **Structural formulae** show which atoms are bonded to each of the carbon atoms in the molecule. They do not usually show any of the C–H or C–C bonds but they give you enough information to work out the structure of the molecule. They are particularly useful to show the structure of longer or more complicated molecules like decane where there can be side chains.

$$CH_3CH_2CH_3$$
propane

$$CH_3CH_2CH_2CH_2CH_2CH_2CH_2CH_2CH_2CH_3$$
decane

The structural formulae of propane and decane.

Alkenes

Alkenes are hydrocarbons that contain one (or more) double bonds between carbon atoms.

There must be at least two carbon atoms in an alkene, so the first two members of the alkene family are ethene (two C atoms) and propene (three C atoms).

The double line in these displayed formulae shows the double bond. In a double bond, two pairs of electrons are shared by the two carbon atoms.

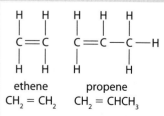

ethene
$$CH_2 = CH_2$$

propene
$$CH_2 = CHCH_3$$

The displayed and structural formulae of ethene and propene.

Naming organic molecules

Organic molecules have names that follow an organised pattern. From the name you can tell how many carbon atoms there are in the chain and also the family the molecule belongs to, as shown in the tables.

First part of name	Number of C atoms in the chain
meth-	1
eth-	2
prop-	3
but-	4
pent-	5
hex-	6

Second part of name	Family
-ane	alkanes
-ene	alkenes
-anol	alcohols (see lesson 5.8)
-anoic acid	carboxylic acids (see lesson 5.8)

Link

You will learn how alkenes are used as the starting materials for the manufacture of polymers in lesson 5.8.

The second part of the name tells you it belongs to the alkene family.

↓

Propene

↑

The first part of the name tells you there are three C atoms in the chain.

Worked example

Draw the displayed formula of pentane.

Step 1 Pentane has five carbon atoms in a chain:

C C C C C

Step 2 Pentane is an alkane, so there are no double bonds between the C atoms. Draw single bonds between the C atoms:

C–C–C–C–C

Step 3 Each carbon atom forms four bonds. Pentane is a hydrocarbon, so all the remaining bonds will be bonds to hydrogen atoms:

Activity B

Use the information in the tables to name:

1 an alkane with three C atoms.

2 an alkene with four C atoms.

Just checking

1 What is the main difference between the structure of an alkane and that of an alkene molecule?

2 Hexene is a hydrocarbon used in the production of some types of polymer. What can you tell about the structure of the molecule hexene from its name?

Activity C

A

$CH_3CH_2CH_2CH_2CH_2CH_3$

B

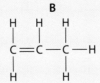

C

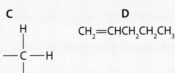

D

$CH_2\!=\!CHCH_2CH_2CH_3$

Which of these structures:

1 are alkenes?

2 is hexane?

3 is an alkane containing a chain of two carbons?

4 is propene?

Lesson outcomes

You should be able to draw and interpret the displayed and structural formulae of alkanes and alkenes, and understand how to represent single and double bonds.

5.8 Other organic molecules

The organic molecules in crude oil are hydrocarbons. But organic molecules from different sources may contain oxygen, chlorine and/or other elements as well.

Get started

Sugars contain carbon, hydrogen and oxygen – for example, glucose has the formula $C_6H_{12}O_6$. Why are sugar molecules not called hydrocarbons?

Key term

Polymerisation – The reaction in which a polymer is formed.

Alcohols

Alcohols have an OH group attached to a carbon atom. Ethanol is the most familiar alcohol. It is made either by the fermentation of sugars or by reacting ethene with water.

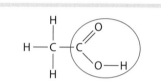

Alcohols like ethanol have an OH group attached to a C atom.

Carboxylic acids

Carboxylic acids have a COOH group at the end of the carbon chain. Although there is still an OH group attached to a C atom, there is also a carbon–oxygen double bond on the same C atom. This completely changes the properties of the molecule, for example its boiling point.

Carboxylic acids like ethanoic acid have a COOH group at the end of the chain.

Activity A

Look at the structures of ethanol and ethanoic acid.

1 List some ways in which the structures are similar.
2 List some ways in which the structures are different.

Chlorohydrocarbons

Chlorohydrocarbons are molecules that contain chlorine atoms as well as carbon and hydrogen.

Remember

It is very important to be able to draw the bonds in these molecules accurately. Each bond should start clearly at one atom and end clearly at another.

Draw double bonds carefully to show that there are two bonds between the atoms. Check that every carbon atom has exactly four bonds to it.

(a) chloroethene   **(b)** dichloromethane

Chlorohydrocarbons, like **(a)** chloroethene and **(b)** dichloromethane, contain chlorine atoms instead of some of the hydrogen atoms.

- The term 'chloro' in the molecule name tells you that there is at least one chlorine atom bonded to a carbon atom, instead of a hydrogen atom.
- You can tell how many chlorine atoms there are from the beginning of the name: di = 2   tri = 3   tetra = 4.

Activity B

Use the information above to draw the structure of tetrachloromethane. Why is tetrachloromethane not called a chlorohydrocarbon?

Polymers

Poly(ethene) is a **polymer** formed when many ethene molecules bond together in a reaction called **polymerisation**. If chloroethene is used in a polymerisation reaction, the polymer is called poly(chloroethene).

(a) poly(ethene)

(b) poly(chloroethene)

Poly(ethene) and poly(chloroethene) have structures which are long chains made up of repeating units.

Although the starting materials are alkenes, there are no double bonds in the final polymer. Polymer molecules are made up of a long chain of carbon atoms bonded together by single bonds.

Link

You will learn more about polymerisation and the uses of polymers in lesson 5.11.

Did you know?

There are two different types of poly(ethene) called high-density poly(ethene) (HDPE) and low-density poly(ethene) (LDPE). They differ in the way in which the polymer chains are arranged and, because of the different arrangements, they have different physical properties and uses. Look out for the names HDPE and LDPE on various types of food packaging.

Assessment activity 5.3 | 2B.P3 | 2B.M3

You are training new technicians working in one of the plants of an oil refinery. The plant separates and purifies the molecules from the gas fraction of crude oil.

1 Prepare a poster that explains how to represent the molecules in different ways – name, structural formula and displayed formula. Include alkanes with 1–6 carbon atoms, as well as ethene and propene.

2 Be able to present this poster and explain what it shows.

Tips

For level 2B.P3, you need to be able to accurately draw displayed formulae for all the molecules.

For 2B.M3, you need to make sure that in your presentation of the poster you correctly explain the full meaning of the single and double lines in the displayed formulae.

A technician working in an oil refinery.

Just checking

1 Draw the displayed formulae of:
   (a) methanol
   (b) propanoic acid
   (c) trichloromethane.

2 Poly(ethene) is a polymer made up of many repeating units bonded together. Draw a section of the polymer chain which contains three repeating units.

Lesson outcome

You should be able to draw and interpret displayed and structural formulae of ethanol, ethanoic acid, chlorohydrocarbons and polymers.

5.9 Identifying hydrocarbon molecules

Get started

How would you test a liquid to show whether it was an acid or an alkali?

Key term

Addition reaction – A reaction in which two molecules join together to make a single product molecule.

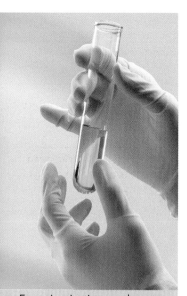

Forensic scientists may have to test unknown liquids to find out what they are.

In the next two lessons, you will learn about some tests that will help you to identify organic molecules.

Identifying unknown chemicals

Forensic scientists analyse evidence brought in from crime scenes. This could be paint chips from cars, blood samples or bottles of unlabelled liquids, perhaps being used in the preparation of illegal substances.

Imagine you are a forensic scientist trying to identify a clear, colourless, organic liquid. Is it an alkane, an alkene, an alcohol or a carboxylic acid?

Using a flowchart is a good way to help you decide how to test your substance to identify it. You will find out more about the tests in this flowchart in the next few pages.

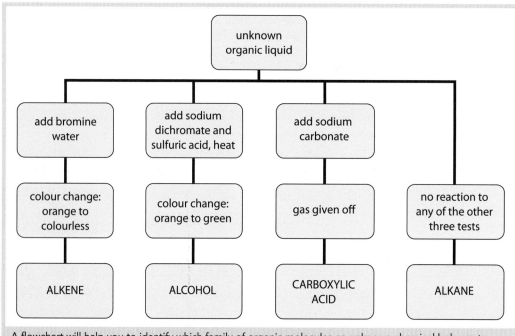

A flowchart will help you to identify which family of organic molecules an unknown chemical belongs to.

Different kinds of tests

A good test must give an obvious positive result – a colour change perhaps, or a gas given off.

Here are two tests that can be done easily in the laboratory. They are carried out on liquid substances.

1 Fill a test tube about a quarter full with water. Add a few drops of the unknown liquid and shake gently.

   If the liquid floats on top, then it is **insoluble** in water. Many organic chemicals are insoluble in water because they cannot form bonds to the water molecules. However, alcohols and carboxylic acids can form bonds to water, so they are **soluble** in water.

Safety and hazards

When you carry out these tests you will often be using pure samples of organic chemicals, not solutions. Many of these substances are flammable and harmful. It is very important to only use a few drops of the chemical when you carry out each test. Make sure you check how to dispose of the chemicals after you have used them.

2 Fill a test tube about a quarter full with bromine water. Add a few drops of the unknown liquid, place a bung on the top of the test tube, and shake gently.

If the bromine water changes colour from orange to colourless then the liquid is an alkene. The bromine has reacted with the double bond in the alkene so its colour disappears.

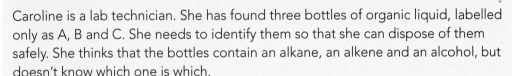
The orange colour of bromine disappears when it reacts with an alkene.

Alkenes react with bromine because they take part in an **addition reaction**. Two molecules react together to form just one product.

Alkenes react with bromine in an addition reaction.

Activity A

Look at the structure of the product which is formed when bromine reacts with ethene. Is this product molecule an alkene? How can you tell?

Activity B

Caroline is a lab technician. She has found three bottles of organic liquid, labelled only as A, B and C. She needs to identify them so that she can dispose of them safely. She thinks that the bottles contain an alkane, an alkene and an alcohol, but doesn't know which one is which.

- Liquid A dissolves in water.
- Liquid B does not dissolve in water. However, the colour of bromine water disappears when a few drops of liquid B are added to it.
- Liquid C does not dissolve in water and does not react with bromine water.

Can you identify which liquid is which?

Just checking

Look at the names of these four organic molecules:
A: hexane (an alkane) **B:** hexene (an alkene) **C:** ethanol (an alcohol) **D:** ethanoic acid (a carboxylic acid).
1 Which molecules are soluble in water?
2 Which molecule will react with bromine, turning it colourless?
3 Which molecule is insoluble in water and will not react with bromine water?

Lesson outcome

You should be able to use test tube reactions to tell the difference between alkanes and alkenes.

5.10 Identifying more organic molecules

Get started

In previous lessons you've come across some families of organic molecules which contain oxygen atoms. Can you name two of the families?

Key terms

Neutralisation – A reaction in which an acid reacts with a base to form a neutral salt and water.

Oxidation – A reaction in which a molecule gains oxygen atoms.

Link

To find out about the colour of universal indicator for different values of pH, see Principles of Applied Science lesson 1.15.

Activity A

What is the gas given off when an acid reacts with a carbonate? How could you test the gas to check this?

Safety and hazards

Sodium dichromate(VI) solution is toxic. You should wear gloves when handling it. When you shake the mixture you must make sure that it does not spill out of the test tube. Placing a bung on the test tube can help.

Is the substance a carboxylic acid?

You will normally use a solution of carboxylic acid because pure carboxylic acids are corrosive.

- Fill a test tube about a quarter full with a solution of a carboxylic acid.
- Add a few drops of universal indicator (UI), which is green to start with.
- An orange–red colour, or a pH of between 2 and 4, shows you that the liquid is an acid. However, there are several other types of organic molecule with acidic properties.
- To check that the substance is a carboxylic acid, add a small spatula measure of sodium carbonate and shake gently. A gas should be given off.

Carboxylic acids are the only common type of organic molecule that reacts with sodium carbonate.

> ethanoic acid + sodium carbonate → sodium ethanoate + carbon dioxide + water

This is a **neutralisation** reaction, and the sodium ethanoate is a salt – just like sodium chloride and calcium sulfate, which you met in Principles of Applied Science Unit 1.

Is the substance an alcohol?

To test if a substance is an alcohol:

- Prepare a water bath by filling a $250\,cm^3$ beaker about half full with hot water from a kettle.
- Add $1–2\,cm^3$ of sodium dichromate(VI) solution to a test tube.
- Then add an equal volume of dilute sulfuric acid.
- Finally, add a few drops of the unknown liquid and shake gently.
- Place this test tube carefully in the hot water bath. Watch what happens over the next few minutes.

Acidified sodium dichromate(VI) solution goes from orange to green when it is warmed with an alcohol.

If the mixture changes colour from orange to green, then the liquid is an alcohol.

Alcohols react with acidified sodium dichromate in an **oxidation** reaction. The alcohol molecule gains an oxygen atom.

Ethanol is oxidised to ethanoic acid by gaining an oxygen atom.

Activity B

Rahul is a quality control chemist. He is monitoring the products of a reaction in which propanol (an alcohol) is being oxidised into propanoic acid (a carboxylic acid). Rahul suspects that in one of the batches, the reaction has not worked. Both the organic substances are colourless liquids so the reaction is not visible.

To check if the reaction has occurred, Rahul carries out two tests on each batch.

First he adds universal indicator to a sample from each batch.

Then he takes a second sample from each batch and heats it with acidified sodium dichromate.

Here are his results.

Batch	Observations when universal indicator (UI) is added	Observations when heated with acidified sodium dichromate
A	UI turns from green to orange	Sodium dichromate stays orange
B	UI stays green	Sodium dichromate turns from orange to green

What can Rahul conclude about the two batches?

Take it further

The oxidation of alcohols to carboxylic acids is an example of a reaction which happens in two separate steps.

Step 1: the alcohol is oxidised to an aldehyde which contains a C=O bond but not an OH group.

Step 2: the aldehyde is oxidised to a carboxylic acid, which contains both C=O and OH.

Assessment activity 5.4 | 2B.P4 | 2B.M4 | 2B.D3

You are the senior technician working in a large chemical company. The company uses large amounts of four substances: hexane, hexene ($CH_2=CHCH_2CH_2CH_2CH_3$), ethanol and ethanoic acid. All of these are clear, colourless liquids. You have found four unlabelled bottles of liquid in the laboratory.

1 Write a report describing suitable tests you could carry out that would enable you to work out what is in each bottle.

2 Your report should include an explanation of what would happen in any reactions that you would carry out.

Tips

For 2B.P4, you should be able to identify an alcohol and a carboxylic acid from your results, as well as an alkane and an alkene.

For 2B.M4, you need to be able to explain how you used your results to make these identifications.

For 2B.D3, you need to be able to explain what happened in each of the reactions, including the structures and bonding of the molecules involved, and to give a correct description of what type of reaction has occurred.

Just checking

1 An unknown organic substance dissolves in water and fizzes when it is added to sodium carbonate. What family of molecules does it belong to?

2 Look at the names of these three substances:
   A: ethanoic acid (a carboxylic acid)
   B: ethanol (an alcohol)
   C: hydroxybutanoic acid (contains an alcohol and a carboxylic acid group).
   (a) Which molecules will turn universal indicator solution from green to orange?
   (b) Which molecules will turn acidified sodium dichromate solution from orange to green when they are heated together?
   (c) Which molecule does not fizz when it is added to sodium carbonate?

Lesson outcome

You should be able to use test tube reactions to identify carboxylic acids and alcohols.

5.11 Fuels, polymers and food products

Get started

Think about everything you have done since you got up this morning. How many of these activities involved using organic chemicals?

Did you know?

Since 2010, petrol companies in the UK have been legally required to include 5% bioethanol in their petrol.

Activity A

1 What are the advantages of using a renewable energy source, like bioethanol, rather than a fossil fuel like pure petrol?

2 Using the Internet, find out some disadvantages of adding bioethanol to petrol used in cars.

Link

You learned about polymers in lesson 5.8.

Fuels

You have already seen that many alkanes are used in fuels, including natural gas, LPG and petrol.

Alcohols, such as ethanol, are also good fuels. They burn in a similar way to alkanes but produce less soot and carbon monoxide, so they are a cleaner, safer fuel.

> ethanol + oxygen → carbon dioxide + water

Bioethanol is ethanol that has been made by **fermenting** the sugar in plants such as sugar cane. Bioethanol is a renewable energy source. Some cars have been designed to run on almost pure ethanol, but even normal petrol engines work if small amounts of ethanol are added to the petrol.

Polymers

When alkenes are heated up at a high pressure and with a catalyst, polymerisation occurs. Different alkenes are used as the **monomers** in the reaction.

Ethene and chloroethene are used to make the polymers poly(ethene) and poly(chloroethene). The double bonds in the alkene molecules break and this allows the monomers to join together to make polymers.

These polymers are long chain molecules containing C–C single bonds.

Chloroethene monomers react to form poly(chloroethene).

Case study

Jenny is a polymer chemist working for a chemical company producing poly(ethene) for specialised uses.

'I have to match the properties of poly(ethene) to the customers' requirements – different uses need different strengths or flexibilities. To change the properties of a polymer we use a process called co-polymerisation – this means using a mixture of monomers, for example 80% ethene and 20% propene. Using propene produces some branches in the polymer chain so the chains don't slide past each other so easily, which means that the polymer is less flexible.'

1 What would you predict about the flexibility of a polymer made from 60% ethene and 40% propene?

Activity B

Write an equation, using words and formulae, to show the polymerisation of ethene to form poly(ethene).

Poly(chloroethene) is commonly known as PVC – polyvinyl chloride, after the old name for chloroethene. There are strong forces between the molecules in the chain, making PVC a strong and rigid material. To make PVC more flexible, small amounts of a molecule called a plasticiser are added. This plasticised PVC is used to make insulating coverings for electrical wiring. The rigid, 'normal' form of PVC is often known as unplasticised PVC (uPVC). It is used in building products such as window frames.

These uPVC window frames will remain strong and rigid even on the hottest day.

Polytetrafluoroethene (PTFE) is a polymer made from a monomer that contains only carbon and fluorine. The PTFE polymer is very slippery – there is no friction when substances rub against it. PTFE can be used to make non-stick cooking utensils and low-friction bearings in machine tools. Its trade name is Teflon™.

Food products

Almost everything we eat is an organic substance because it comes from something that was once alive. Most of the reactions that produced these substances happened in the organism itself – such as photosynthesis which produces sugars in plants.

Food will not stick to the PTFE surface of this frying pan.

However, some food products are made by chemical reactions controlled by industrial chemists. Ethanol, produced by fermentation, is present in alcoholic drinks. Ethanol changes the taste of the drink, helps to preserve it and acts as a drug to anyone who drinks it.

If ethanol is left in contact with the oxygen in the air, it can be oxidised to ethanoic acid. This carboxylic acid is used in vinegar to give foods a 'sharp' taste. Ethanoic acid also helps to preserve foods because bacteria cannot live in acidic conditions.

Just checking

1 Look at this list of organic molecules and their uses. For each use, explain what property of the molecule allows it to be used in this way.
   A Ethanoic acid is used to preserve foods.
   B PVC is used to cover electrical wires.
   C Ethanol is used as an alternative to petrol in car engines.
   D PTFE is used to coat frying pans.

Lesson outcome

You should be able to give examples of the use of organic molecules in fuels, as polymers, and in foodstuffs.

Get started

How would you test a substance to see if it dissolved in a particular solvent, such as water or ethanol?

Key terms

Esters – Organic molecules that contain the functional group –COO–.

Solvent – A liquid in which other substances (called solutes) will dissolve to make a solution.

Did you know?

Ethanol can be used as a sterilising agent to kill microorganisms. Hand wipes and sterilising gel contain ethanol.

Safety and hazards

Dichloromethane is toxic and its use as a paint stripper in homes was banned by the EU in June 2010. The use of DCM in industrial installations is still permitted.

Solvents

Ionic substances (such as salts) dissolve in water. Water is a good **solvent** for these substances.

Most organic substances do not dissolve in water. To make a **solution** from these substances you need a different solvent – another organic compound.

Ethanol is used as a solvent in several different types of product.

- Perfumes are made from very concentrated scent molecules dissolved in ethanol to make a dilute solution. When you dab perfume or aftershave on your skin, the ethanol evaporates to leave the scented molecules behind.

Perfume is a solution of scent molecules in ethanol.

- Ethanol is used in cosmetic products, for example hair spray. The ethanol evaporates to leave a hard lacquer coating the hair.

- Many permanent inks are solutions of coloured dyes dissolved in ethanol.

Activity A

Suggest two properties of ethanol, apart from the fact that it is a good solvent, which are important for its use in perfumes.

Another organic molecule used as a solvent is dichloromethane (DCM), which can be used as a paint stripper. A gel containing DCM is placed on the paint surface. The DCM starts to dissolve the paint, which detaches itself from the surface. The paint can then be scraped off.

DCM is also a good solvent for the caffeine molecules found in coffee beans. One way of making decaffeinated coffee is to rinse the beans with DCM for several hours. The beans must then be carefully washed to get rid of any traces of the DCM solvent.

Feedstocks

You have already seen how ethene is used as a **feedstock** (starting material) to manufacture polymers. Ethene is also used in industry in a reaction with steam, which produces large amounts of ethanol very rapidly. A temperature of 300 °C, a phosphoric acid **catalyst** and high pressure are needed for this reaction.

Ethanol is produced industrially by reacting ethene with steam.

The process of fermentation uses yeast to break down sugars to produce ethanol and carbon dioxide. A temperature of 30–40 °C and normal pressure are needed for this reaction. The enzymes in the yeast act as the catalyst, but the reaction is slow.

The equation for the reaction is:

$$sugar \rightarrow ethanol + carbon\ dioxide$$
$$C_6H_{12}O_6 \rightarrow 2C_2H_5OH + 2CO_2$$

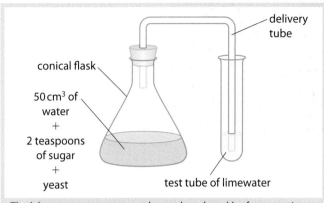

The laboratory apparatus used to make ethanol by fermentation.

Activity B

Compare the two methods of making ethanol – by fermentation or by reaction with steam. Think of one advantage and one disadvantage of each method.

Ethanoic acid is used as a feedstock in a reaction with ethanol or other alcohols to produce molecules called **esters**.

The equation for reaction which produces esters.

Esters have fruity smells and are used in perfumes and food flavourings. Fruits like pears, bananas, oranges, raspberries and pineapples, contain natural esters that give them their distinctive smell. Food technologists use both natural and artificial esters to create flavourings for foods that can fool all but the most professional tasters.

Activity C

Look back over the last two lessons, 5.10 and 5.11, and then list all the ways in which ethanol is used.

Just checking

For each of the four molecules **A–D**, select its most likely use from the list below.
A: ethene **B:** ethanoic acid **C:** dichloromethane **D:** ethyl ethanoate (an ester).

1 A food flavouring.
2 A solvent for decaffeinating coffee.
3 A feedstock for making polymers.
4 A feedstock for making esters.

Lesson outcome

You should be able to give examples of the use of organic molecules as solvents and feedstocks.

5.13 Problems with organic molecules

Get started

Write a list of some properties of organic substances that make them useful.

Key term

Biodegradable – Able to be broken down by the action of living organisms such as bacteria.

Discussion point

On average, every person in the UK produces about 50 kg of plastic waste every year. Research the different ways of disposing of this waste, and discuss the advantages and disadvantages of these methods.

You have seen how organic molecules have properties that allow them to be used in a variety of ways, such as solvents and fuels. However, organic molecules often have other properties that can cause problems when they are used. Organic chemists need to decide whether the benefits of using the organic molecule outweigh the drawbacks.

What are the drawbacks?

- Flammability: most organic molecules are **flammable** (catch fire easily). If an organic substance has a low boiling point, this can create a serious fire risk when the vapours build up in an enclosed space.
- Toxicity: some organic molecules are **toxic** if they are breathed in or absorbed through the skin. This means that precautions need to be taken when using or disposing of them.
- Disposal in landfill: if organic molecules are toxic they will cause environmental problems if they leach into the soil, lakes or rivers. Some organic substances are not **biodegradable**, so they will last for thousands of years in landfill sites.
- Disposal by burning: this produces greenhouse gases like carbon dioxide and can also release new toxic products into the atmosphere.
- Non-renewable: most organic molecules are made using crude oil as a raw material. This is non-renewable and so is gradually running out.

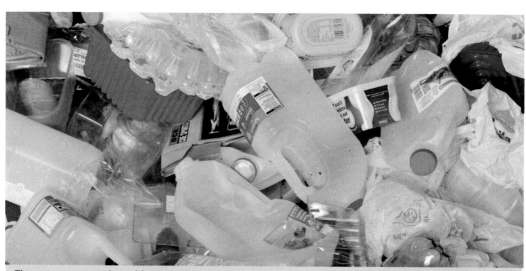

There are concerns that additives in this packaging could leak into the environment.

Activity A

Greenhouse gases are molecules that trap energy which is emitted by the Earth, in the form of infrared radiation. This means that, as the amount of greenhouse gases in the atmosphere increases, the temperature of the atmosphere will increase as well. All industries now need to demonstrate that they are working to reduce their emissions of greenhouse gases.

Research, using books or the Internet, at least three different greenhouse gases and the steps being taken to reduce their emission.

Polyvinyl chloride (PVC)

PVC is used as a building material, for example in window frames and water pipes, as well as for insulation for cables, plastic coverings for furniture and packaging. Like most polymers, PVC is not toxic, but there are a number of drawbacks to using it. These have to be weighed against the benefits, as you will see summarised in the table.

Benefits of PVC	Drawbacks of PVC
Low cost	Plasticised PVC contains plasticiser molecules, called phthalates. These are slowly released from the PVC into the environment. They can affect the working of hormones in the body and could cause birth defects and illness in children.
Can be made rigid or flexible as required	When PVC is burnt, it releases molecules called dioxins, which are similar to some of the chemicals used in weedkillers. They are toxic molecules, and can build up in the food chain and can cause skin diseases and birth defects.
Weatherproof (not affected by water)	PVC molecules are not biodegradable so they will remain in landfill for thousands or millions of years.
PVC molecules are non-toxic	PVC is made from crude oil, which is a non-renewable resource.

Case study

Rupinder works for a company which runs a large incinerator.

'My job is to monitor the emissions produced by the incinerator. When the company built this incinerator it was very controversial – local residents were very worried by the possibility of hazardous emissions threatening their health. Every day I collect data on the emissions produced and we publish this on our website. There are very strict limits on the emissions that can be produced and we are well below these for substances such as particulates and volatile organic compounds, which can include dioxins. The nitrogen oxide emissions are the only ones which get anywhere near the government's limit of $200\,mg/m^3$.'

1 Why does Rupinder specifically mention dioxins?

2 Find out some of the environmental effects of nitrogen oxides. What are the other main sources of nitrogen oxides?

Just checking

1 List two ways of disposing of organic molecules.
2 What environmental problems are caused when organic molecules are disposed of in these ways?
3 Why should people be particularly careful when handling and using organic molecules?

Link

You learned about polymers in lesson 5.8 and about PVC in lesson 5.11.

Did you know?

One of the biggest uses for PVC in the past was the manufacture of 'vinyl' records. Some musicians and DJs still prefer this way of recording and playing music, and the records they use are still made from PVC.

Lesson outcome

You should be able to explain some of the problems associated with the use of organic molecules.

5.14 Benefits and drawbacks

Activity A

Look back at the information about PVC in the previous lesson. Do you think the advantages outweigh the disadvantages, for each of these uses?

1 Using unplasticised PVC to make underground water pipes.
2 Making a child's toy from plasticised PVC.

Various organisations and people may have to decide whether the benefits of using an organic chemical outweigh the risks.

- Governments may have to consider whether to pass laws to ban or restrict its use.
- Consumers may read stories in the media about the risks. They have to decide whether they are prepared to take the risks of using it.
- Manufacturers will need to carry out research to find out whether they can make a product less **hazardous** or find alternative substances to use.

Do the advantages of using this plasticised PVC outweigh the drawbacks?

Assessment activity 5.5　　| 2B.P5 | 2B.P3 [part] | 2B.M5 | 2B.D4

You are a development chemist working for a company that manufactures a range of chemicals for consumer products. These include: PVC from chloroethene, poly(ethene) from ethene, ethanol from ethene, ethanoic acid from ethanol and a range of chloromethanes.

The marketing team are concerned that reports of problems with the use and disposal of some of these molecules is affecting sales. They have asked you to write a brief report on the uses of these molecules, and the problems involved, focusing particularly on PVC and dichloromethane.

Tips

You need to identify the uses of ethene, ethanol and ethanoic acid.

For 2B.P3, your report should also include drawings of the structures of all of these molecules.

For 2B.P5, you should include a brief description of one use of each of these molecules.

For 2B.M5, you should explain in detail the problems and risks involved in the use or disposal of PVC and dichloromethane.

For 2B.D4, you must discuss the benefits of these molecules in more detail, and reach a judgement about whether the benefits outweigh the risks and problems.

Lesson outcome

You should be able to evaluate the use of some organic molecules, considering the benefits and drawbacks.

WorkSpace

Pierre
Environmental Health Officer

As part of my work, I am the chair of a committee that looks at the hazards involved in the use of various chemicals. One of the substances we have looked at is dichloromethane (DCM). It is a really good solvent for a variety of organic molecules, including some of the components of paint.

A lot of chlorohydrocarbon molecules are very toxic, but DCM is the least toxic. So, because of this, and the fact that it can be made cheaply, it has been really popular in the last few decades as an ingredient in paint stripper. But it is still toxic if it is breathed in, and because it has a low boiling point it can build up quickly in enclosed spaces.

The EU has banned this substance for use in the home, but we had to decide whether to request an exemption for use by professional decorators. The clincher for me was that some recent studies have suggested that there might be an increased risk of cancer among professional decorators who are using it all the time.

Because there are other ways of removing paint – maybe a combination of heat-stripping and using other chemicals – we decided that the drawbacks outweighed the benefits for this substance, and we wrote a report recommending that it should continue to be banned in the home, even when used by professionals.

Think about it

1 The committee didn't ban other uses of DCM, such as decaffeinating coffee. Why not?

2 Why do you think it has taken until now to discover that DCM may cause cancer?

3 Draw up a table, like the one for PVC in lesson 5.13, to summarise the benefits and drawbacks of using dichloromethane.

5.15 Nanochemicals

Get started

Look at this table.

Particle	Diameter
Grain of sand	1 000 000 nm
Human hair	100 000 nm
Nanoparticle	Less than 100 nm
Carbon atom	0.3 nm

1 nm (nanometre) is one billionth of a metre (10^{-9} metres).

The largest nanoparticle is almost 100 nanometres in diameter. How many times smaller is it than a grain of sand? How many times bigger is it than a carbon atom?

Key term

Nanochemical – A substance made up of nanoparticles.

Link

The photograph on the opening page of this unit shows a scanning electron micrograph of a nanowire.

Did you know?

Architects have known for many years that buildings can be made very strong by creating a latticework of hexagons and pentagons. Fullerenes are named after a man called Buckminster Fuller – an architect who designed domes looking very like nanospheres. The first carbon nanosphere was called Buckminsterfullerene or the 'buckyball'.

Nanochemicals are substances that are made up of **nanoparticles**. Nanoparticles may be molecules or small crystals containing only a few tens or a few hundreds of atoms. They can also be thin sheets or tiny tubes.

At this scale, called the nanoscale, chemicals can behave in very different ways to the way they normally behave. Chemists working in this branch of science, called nanochemistry, are discovering many exciting applications of nanochemicals.

Carbon nanostructures

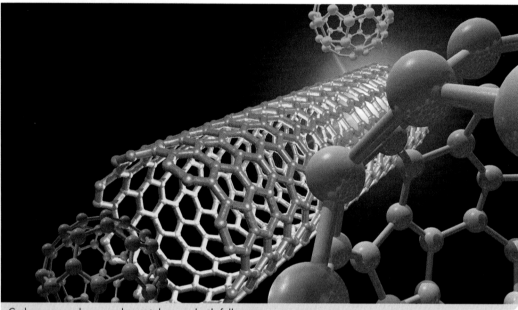

Carbon nanospheres and nanotubes are both fullerenes.

In 1985, scientists found that carbon atoms could arrange themselves into nanospheres or nanotubes. They called these structures fullerenes. Fullerenes are very strong and stable because of their structure. Their carbon atoms are arranged in the form of hexagons and pentagons, which are very stable arrangements.

Nanotubes can be produced by putting hydrocarbon molecules into contact with a very hot catalyst containing metal nanoparticles. The carbon atoms from the hydrocarbon attach themselves to the metal particles and a tube starts growing from them.

Carbon nanotubes can be used in a variety of applications, for example:

- in specialised clothing: the nanotubes are very flexible but also strong and light; they can be used in protective sports gear and knife-proof vests
- to add strength and flexibility to sports equipment, such as tennis racquets
- electrodes in batteries for mobile phones: they **conduct** electricity but have a very high surface area, which means the battery can be recharged faster.

Nanowires are single crystals of carbon that conduct electricity, such as graphite, but are very thin and light. Other single-crystal nanowires have been made from substances such as platinum or silicon oxide. Nanowires have a lower conductivity than carbon nanotubes and may be used in the future in microprocessors and other electronic devices.

Nanotubes are being used to replace metals in electrical circuits because they have better properties. Use ideas about the structure of nanotubes to explain why:

1 a carbon nanotube will be less dense than a copper wire.

2 electrodes made from carbon nanotubes will have a greater surface area than an electrode made from a piece of steel.

Graphene

Graphene is a third type of nanochemical based on carbon. It is a flat sheet of carbon atoms that is just one carbon atom thick, and so it is called a nanosheet.

Graphene can be made from graphite, which has a structure made up of many sheets of carbon atoms stacked on top of each other. It was only in 2004 that a pure and stable sample of graphene was produced.

Graphene has several exciting properties.

- It is incredibly strong – 200 times stronger than steel – but also flexible.

- It is transparent.

- It acts as a semiconductor.

- It is chemically very stable.

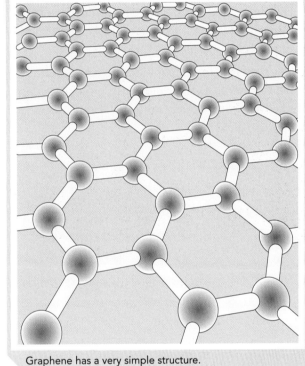

Graphene has a very simple structure.

? Did you know?

Although scientists have tried many complex ways of producing graphene, the method that works best is amazingly simple – they just used pieces of sticky tape to pull the layers of graphite apart.

Because it is so new, scientists can only guess at how it may eventually be used. Some possibilities are:

- ultra-thin coatings to protect objects from attack by chemicals

- as a material for building lightweight cars or planes

or even…

- flexible tablet computers that you could fold up and keep in your pocket.

Just checking

1 What is a nanochemical?

2 Name two kinds of carbon-based nanochemicals.

3 Give two uses of nanowires.

Lesson outcomes

You should know the meaning of the term nanochemical and be able to give examples of uses of nanotubes.

5.16 Uses of nanochemicals

Get started

Can you think of any nanochemicals which you might use in your everyday life? What properties of nanoparticles make them useful?

In the last lesson, you saw how carbon nanotubes and nanosheets are starting to be used in a variety of ways, such as specialised clothing, sports equipment and electrical components.

Most of the nanochemicals being used at the moment are nanospheres – either made from carbon or from other elements or compounds.

Carbon nanospheres

Carbon nanospheres are incredibly strong and stable because of the way the atoms are arranged. Just like a football, even if the sphere is squashed out of shape, it will spring back into a spherical shape again afterwards.

Did you know?

Carbon nanospheres are so strong that they can withstand being slammed against a steel plate at speeds of over 15 000 mph.

- These strong nanospheres can be used as cages to trap molecules inside. Hydrogen gas, used as a fuel, could be stored inside a nanosphere. The gas molecules would be squashed so close together that it would have the density of a solid, not a gas.
- Alternatively, organic molecules can be attached to the outside of the sphere. Nanospheres coated with drug molecules can penetrate cells or bind to receptor sites on the cell membrane and be used to treat cancer and HIV, the virus that causes AIDS.

Activity A

Why do you think it is an advantage to be able to increase the density of hydrogen gas when storing it as fuel?

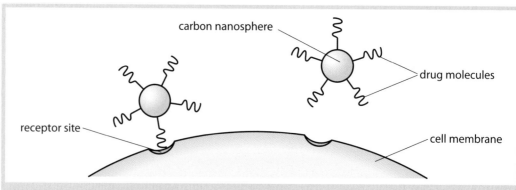

Drug molecules attached to nanospheres can bind to receptor sites on cell membranes.

Activity B

The table below shows some of the properties of nanospheres used in drug delivery systems and some of the advantages of using these systems.

Try and match up the property with the advantage which it produces.

Property of nanospheres	Advantage of using nanospheres
Nanospheres have a large surface area.	The drugs delivered by nanospheres are only slowly broken down by the body.
Molecules attached to nanospheres are very stable.	Drugs can be taken by swallowing pills or as nasal sprays.
Nanoparticles can pass into the body in a variety of ways.	Many drug molecules can be attached to one nanosphere.

Other nanospheres

Nanospheres can by made from other substances apart from carbon.

Zinc oxide and titanium dioxide are both white solids. They are used in sun creams because they reflect harmful ultraviolet radiation very well.

If nanocrystals of these metal oxides are used, they still reflect ultraviolet radiation but, because the crystals are so small, they do not reflect visible light. This means that the sun cream does not look white against the skin.

Nanoparticles of other coloured minerals – ceramics or even volcanic ash – are used in cosmetics like nail varnish or mascara. The nanoparticles mean that the cosmetic is easier to apply and forms a thinner, more even layer on the skin.

Plastic bags and plastic bottles can be made extremely strong by mixing the plastic with clay or carbon nanoparticles. The nanoparticles help to strongly 'glue' together the polymer molecules in the plastic. This **composite** material is now strong, flexible and does not allow gas molecules to pass through it, which is useful for storing fizzy drinks.

Drug molecules can be trapped within a nanosphere made from a polymer. Once inside a cell, the polymer breaks down, releasing exactly the right dose of the drug to the target.

Mascara made from nanoparticles is easier to apply evenly.

Activity C

You are a science journalist working for a TV news website. Using the Internet and information from the last two lessons, research the use of nanoparticles in one of these key areas:

- sports equipment
- food
- clothing
- electronics
- medicine.

Write a short article for the website outlining all the different uses you have found, and explain the advantages of using nanoparticles.

Just checking

Look at these uses of nanoparticles. Each of these uses depends on a particular property of the nanoparticles.

Complete each of the sentences to explain why nanoparticles can be used in this way (all the information you need can be found in this lesson). The first sentence has been completed for you.

1 Nanoparticles can be used in drug delivery systems because...*they are small enough to penetrate cells.*
2 Nanoparticles can be used to store hydrogen because....
3 Nanoparticles can be used to make sun creams invisible on the skin because...
4 Nanoparticles can be used to make plastic bags stronger because...

Lesson outcome

You should be able to give examples of the uses of nanochemicals.

Get started

Some people are very concerned about the use of nanoparticles. Why do you think they are more worried about the use of nanoparticles than about the use of other organic substances such as polymers?

Key term

Ethical issue – A problem for discussion that concerns whether a particular action or decision is right or wrong.

Did you know?

A few scientists have claimed that nanoparticles could bring about the end of life on Earth. This is because they think that certain nanoparticles could start making copies of themselves using carbon atoms from organic substances. This could lead to an uncontrollable chain reaction destroying all organic matter on Earth by converting it into a 'grey goo'. Most scientists today think this is very unlikely!

Nanochemistry is a new and fast-moving branch of chemistry. Research chemists are developing new nanochemicals all the time, and an important part of their job is to assess any possible harmful effects of using these new substances.

Effects on human health

Because nanochemicals are so very small, they may be able to pass easily into the human body.

- If nanoparticles are breathed in they can pass into the tiny airways deep inside the lungs.
- Nanoparticles on the skin could pass through it and enter the bloodstream.
- Nanoparticles in food or drink could be absorbed into the bloodstream as the food is digested.

Once inside the body, the nanochemicals might pass into the cells through the cell membrane. They could then affect chemical reactions inside the cell.

This could be a good thing because nanochemicals can be used to carry drug molecules into cells. But some people claim nanochemicals may cause asthma or even cancer.

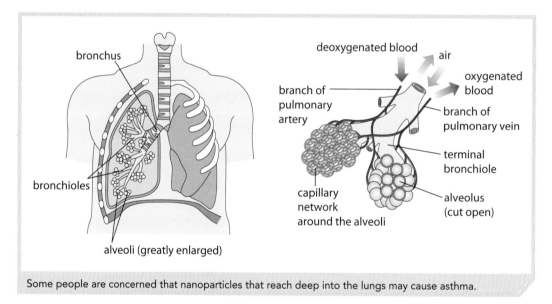

Some people are concerned that nanoparticles that reach deep into the lungs may cause asthma.

Environmental effects

Nanochemicals may enter the environment when they are disposed of.

- When cosmetics and sun creams are washed off they may end up in rivers and lakes.
- Food packaging might be burned (releasing nanoparticles into the air) or dumped in landfill (releasing nanoparticles into the soil).

Nanoparticles could affect animals, insects and fish in the same ways as humans. Nanoparticles may also build up in the food chain, increasing these risks.

Assessing the risk

Some people believe that we do not know enough about the properties of nanochemicals to be sure about the risks. Deciding whether nanochemicals should be widely used is an **ethical issue**.

If you use the Internet to research scientific issues you must check that your source is reliable. This means that it should be based on real scientific evidence and not a **biased source**.

These are the things you should think about and ask yourself when you assess the reliability of an Internet source.

- Find out about the website. Is it of an academic institution, e.g. a university or a scientific body, or is it of an organisation such as a pressure group?
- Does the source explain where it got the information? If it gives details of some scientific research then it may be more likely to be reliable.
- What kind of language does the source use? If it is designed to shock or persuade then it may be biased.

Would you use a sun cream containing nanoparticles?

Activity A

1 Construct a table showing the benefits and risks of using nanoparticles in sun cream.
2 Nanoparticles are now found in many consumer products. Do you agree that the benefits of using nanoparticles outweigh the risks?

Measuring toxicity

When new chemical substances are developed, such as nanochemicals or the drugs that they carry, scientists will try and find out whether the chemicals are toxic

They usually do this by carrying out tests on animals, to see whether large doses of the substance cause illness or death; one way of measuring the toxicity of a chemical is to find the dose of the substance which causes the death of 50% of the animals in a sample. The smaller the dose needed to do this, the more toxic the substance is.

Some people feel that animal tests like this are cruel and **unethical**, but without these tests scientists would not be able to make judgements about how safe a substance is to use.

Other possible effects of chemicals might need animals to be studied over much longer periods, for example to find out whether the substance causes cancer or birth defects. Some people feel that these kinds of tests should be carried out for longer and on a wider range of animals before deciding whether a chemical is safe to use.

 Discussion point

Do you think that using animals to test the safety of chemicals is justified?

 Lesson outcomes

You should know about some of the safety and environmental implications of the use of nanochemicals, and be able to discuss the ethics of using nanochemicals in a range of situations.

Just checking

1 Explain why nanoparticles might cause health risks.
2 Why is the use of nanochemicals an ethical issue?

Designer polymers

Polymer chemists can develop polymers so that they have exactly the right physical and chemical properties for a specialised use.

Kevlar® is a good example of a designer polymer. It was designed to be:

- very strong
- flexible
- lightweight
- fire-resistant.

Kevlar® is used to make bullet-proof and knife-proof helmets and clothing.

Polytetrafluoroethene (PTFE) is used to make a waterproof, breathable fabric called Gore-Tex®. There are tiny holes in the structure of PTFE that are too small for drops of liquid water to pass through from the outside. But the PTFE is described as breathable because molecules of water vapour can pass out through these holes. This makes it much more comfortable for people like hill walkers or runners, who might be sweating inside their clothing or footwear.

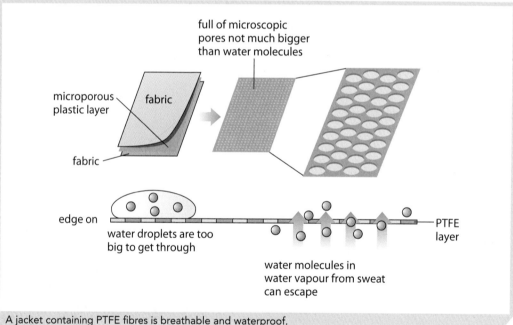

A jacket containing PTFE fibres is breathable and waterproof.

Polyester fleece jackets help you stay warm because they trap air between the fibres. The air is an **insulator**. In Thinsulate® materials, the fibres are very thin, so there is a greater density of fibres in a given area. This helps to trap more air. So even a thin layer will keep you warm on a cold day.

Firefighters need clothing that can insulate them against the intense heat of a fire. Polybenzimidazole (PBI) has an ideal range of properties:

- good thermal insulator
- stable at high temperatures, and non-flammable
- very high melting point
- remains flexible and strong at high temperatures.

PTFE has other uses, completely unconnected with its breathable properties. Use the Internet to find out about other uses.

Smart materials

A **smart material** has properties that change according to the conditions it is in. The table gives some examples.

Type of material	Property	Uses
Shape-memory polymer	The polymer can be bent into different shapes but returns to the original shape when it is heated	Surgical implants such as stents in arteries
Piezoelectric material	Produces an electric current when a force is applied	Sensitive balances
Photochromic material	Changes colour when exposed to light	'Reactolite' sunglasses, which become darker in bright light

The chemicals in this thermochromic thermometer change colour when the child's temperature changes.

Drawbacks of these new materials

Many of these new materials are polymers. There are some drawbacks to making, using and disposing of polymers.

- The feedstocks come from crude oil, which is a non-renewable resource.
- They may contain toxic additives which can leak out of the material.
- They are non-biodegradable and last for thousands of years in landfill.
- If they are burned they can release toxic products.

Assessment activity 5.6
2C.P6 | 2C.M6 | 2C.D5

You are a marketing manager in a company that manufactures a wide range of chemicals. You are looking for new markets for your chemicals, and have been asked to recommend three new substances to produce and market in the next 10 years.

Write a report recommending three suitable substances, including a nanochemical, a material with specialised properties and a smart material.

You should name the substances, explain how they could be used and give the possible drawbacks of using them.

Just checking

Match the four types of molecules **A–D** with the descriptions below.
A Shape memory polymer   **B** Thinsulate®   **C** Kevlar®   **D** PTFE®
1 Contains many layers of trapped air to reduce thermal conductivity.
2 Used in protective clothing because it is strong and lightweight.
3 Can change shape when heated but always returns to its original shape.
4 Contains tiny holes which allow water vapour through, but not liquid water.

Link

Look back to lesson 5.13 to remind yourself of other drawbacks of using polymers.

Tips

You need to explain what nanochemicals are.

For 2C.P6, you will need to describe in detail the uses of your three chemicals, including whether they will be used by themselves or in combination with other materials.

For 2C.M6, you will need to explain the benefits of using these substances in these ways.

For 2C.D5, you will need to compare the benefits and drawbacks of using these substances. This should include information from sources you have researched yourself. You should evaluate and reach a judgement on the reliability of these sources.

Lesson outcomes

You should be able to give examples and understand the uses of specialised and smart materials, and know of some drawbacks.

Introduction

All the physics in this unit has everyday uses. You will start by investigating motion. What is the difference between speed and velocity? The answer explains why cars have to brake as they go round corners. Next you will look at the principle of conservation of energy, and learn why this is essential for designing a good rollercoaster ride.

In the lessons on forces you will learn, for example, what makes a rocket lift off from the ground, and how to work out safe loading levels for a crane. The lessons on optics cover all kinds of equipment and measuring instruments, from glasses to improve your eyesight to telescopes for investigating distant galaxies.

In the lessons on sound you will learn the basis of some sophisticated technology, for example the sonar systems used on ships, voice recognition systems used in telephone networks, and the ultrasound for producing pictures of a growing fetus in the womb. The final lessons look at electrical circuits. You will build a circuit that can be used to control temperature, and find out how to make garden lights that only come on when it is dark.

Assessment: You will be assessed using a series of internally assessed assignments.

Learning aims

After completing this unit you should have:
a investigated motion
b investigated forces
c investigated light and sound waves
d investigated electricity.

What I really enjoyed about this unit has got to be the fact that all the physics we use is related to the real world. I really enjoyed the electrical section where we got to build circuits and find out how thermistors and LDRs are used in home heating systems and garden lights. This will help me when I go on to study electronic engineering.

Callum, *16 year old aspiring engineer*

Applications of
Physical Science

BTEC
Assessment Zone

This table shows you what you must do in order to achieve a Level 1 Pass, or a Level 2 Pass, Merit or Distinction grade, and where you can find activities in this book to help you.

Assessment criteria

To achieve a Level 1 Pass grade, the evidence must show that you are able to:	To achieve a Level 2 Pass grade, the evidence must show that you are able to:	To achieve a Level 2 Merit grade, the evidence must show that you are able to:	To achieve a Level 2 Distinction grade, the evidence must show that you are able to:
Learning aim A: Investigate motion			
1A.1 Produce accurate graphs to represent uniform motion using primary data. Assessment activity 6.1 Assessment activity 6.2	**2A.P1** Produce accurate graphs to represent uniform and non-uniform motion using primary data. Assessment activity 6.1 [part] Assessment activity 6.2	**2A.M1** Interpret graphs to identify objects that are stationary, moving at a constant speed and moving with increasing or decreasing speed. Assessment activity 6.1 [part] Assessment activity 6.2	**2A.D1** Calculate the gradient for distance–time graphs and the gradient and area under speed–time graphs. Assessment activity 6.1 [part] Assessment activity 6.2 [part]
1A.2 Measure distance for simple experiments. Assessment activity 6.1 Assessment activity 6.2	**2A.P2** Calculate speed and velocity for simple experiments. Assessment activity 6.1		
1A.3 Draw energy transformation diagrams for simple experiments. Assessment activity 6.3	**2A.P3** Describe the conservation of energy for simple experiments, including energy transformation diagrams. Assessment activity 6.3	**2A.M2** Calculate kinetic energy and changes in gravitational potential energy. Assessment activity 6.3	**2A.D2** Explain how changes in energy will affect transportation and stopping distances. Assessment activity 6.3
Learning aim B: Investigate forces			
1B.4 Identify the forces on objects. Assessment activity 6.4	**2B.P4** Describe the effects of balanced and unbalanced forces on objects. Assessment activity 6.4	**2B.M3** Calculate the force on objects, in relation to their mass and acceleration for an application. Assessment activity 6.6	**2B.D3** Explain the various forces involved, and their approximate sizes, in a variety of applications. Assessment activity 6.4 Assessment activity 6.6
1B.5 Describe work done in terms of forces moving through a distance. Assessment activity 6.7	**2B.P5** Calculate the work done by forces acting on objects for simple experiments. Assessment activity 6.7		
1B.6 Identify friction forces and situations where they occur. Assessment activity 6.5	**2B.P6** Describe how friction and normal reaction forces are produced in response to an applied force. Assessment activity 6.5	**2B.M4** Explain how friction and normal reaction forces are produced in response to an applied force. Assessment activity 6.5	

Assessment criteria

Learning aim C: Investigate light and sound waves

1C.7	2C.P7	2C.M5	2C.D4
Describe, using diagrams, reflection of light in plane mirrors for simple applications. Assessment activity 6.8	Describe, using diagrams, reflection and refraction of light for simple applications. Assessment activity 6.8 [part] Assessment activity 6.9 [part] Assessment activity 6.10 [part]	Describe how lenses and mirrors can affect rays of light. Assessment activity 6.8 [part] Assessment activity 6.9 [part]	Explain how reflection and refraction of light can be used in applications. Assessment activity 6.8 [part] Assessment activity 6.10 [part]
1C.8 Describe how sound is reflected for simple applications. Assessment activity 6.12	**2C.P8** Describe the importance of a medium for the transmission of sound waves through a variety of substances for simple applications. Assessment activity 6.11 Assessment activity 6.12	**2C.M6** Describe the propagation of sound waves, including compression and rarefaction. Assessment activity 6.11	**2C.D5** Explain how sound waves can be applied in everyday uses. Assessment activity 6.12

Learning aim D: Investigate electricity

1D.9	2D.P9	2D.M7	2D.D6
Describe, using diagrams, how to build series and parallel circuits. Assessment activity 6.13	Measure currents and voltages in series and parallel electric circuits. Assessment activity 6.13	Calculate resistances from measured currents and voltages. Assessment activity 6.13	Analyse an everyday life situation in which the resistance of a conducting wire is not constant. Assessment activity 6.14
1D.10 Describe the use of a thermistor or LDR for an application. Assessment activity 6.15	**2D.P10** Investigate an application of thermistors or LDRs using primary data. Assessment activity 6.15	**2D.M8** Mathematically or graphically process the results of the investigation into thermistors or LDRs to draw conclusions. Assessment activity 6.15	**2D.D7** Evaluate the investigation into thermistors or LDRs, suggesting improvements to a real-life application. Assessment activity 6.15

How you will be assessed

The unit will be assessed by a series of internally assessed tasks. You will be expected to carry out a number of practical investigations involving forces, motion, waves and electricity. You will develop your practical skills and will be expected to interpret experimental results using knowledge that you will acquire in this unit.

Your assessment could be in the form of:

* written observation sheets from your teacher, stating that you carried out the experiments safely and correctly
* experimental reports, giving details of your experiments
* poster presentations, showing the results of your investigations.

Within each lesson in this unit there are also activities designed to make you think and research further into some of the topics.

6.1 Investigating motion

Get started

Discuss, in groups of three, a recent car or bus journey that you made. In your discussion, think about the following questions: Do you know how far you travelled? How would you know the distance? Is there anything in the car or bus that would tell you the journey distance? Did you notice any changes to the speed of the vehicle during the journey?

Key terms

Displacement – The shortest distance from the initial to the final position of an object that has moved.

Motion – Movement that results in an object changing its position.

Non-uniform motion – An object moving with velocity or acceleration that is changing.

Uniform motion – An object moving with velocity or acceleration that is not changing.

When you move from one position to another, for example, walking from your house to the shop, your movement is called **motion**. If you travel at the same speed all the time, it is called **uniform motion**. If you change speed during your journey, for example if you slow down going up a hill or stop to chat to a friend, it is called **non-uniform motion**. Most real-life situations will involve non-uniform motion because the moving object will be speeding up and slowing down.

◤ Measuring distance

Distance is the length of the path an object travels as it moves from one point to another. For example, on a 400 m running track, the distance covered in one lap is 400 m. The **displacement** is the distance measured in a straight line from a reference point. So, if the reference point is the starting line of the 400 m lap (which is also the finish line), the displacement is zero, although the distance covered in one lap is 400 m.

Activity A

One lap of the athletics track shown below is 400 m.

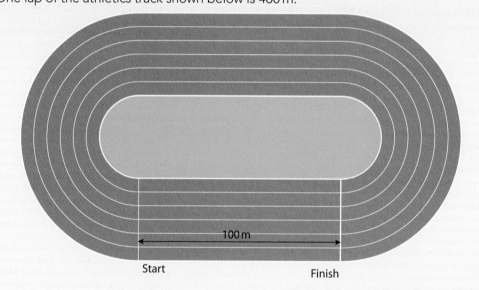

A race is six laps of the track, plus a short extra section, the distance between the start and the finish lines. Calculate:

1 the total race distance
2 the displacement over the whole race.

In the real world, there are many situations in which people need to measure distance. For example, in a road traffic collision, investigators use tape measures and **measuring wheels** to measure skid marks on the road, and the distance between the first marks and the collision site. They can use this information to work out how fast the vehicles were going when the accident happened. Accident investigators also use laser measuring devices that give more accurate measurements.

In the laboratory, we usually measure distance using rulers or tape measures. Distances can also be measured by electronic devices that use a beam of infrared radiation (similar to a TV remote control).

Measuring distance using an infrared beam of radiation.

Measuring time

One place where time measurement is very important is in sports events such as the 400 m. Good sports stopwatches measure hundredths of a second. The timers used in international sporting events are accurate to one thousandth of a second.

In the laboratory you may use a stopclock or stopwatch to measure time intervals. If you are investigating **motion**, you may use **light gates** to measure the time it takes for a trolley to move between two points.

The card on the trolley cuts a light beam as it passes through each light gate. The data logger records the time at which each light beam is cut.

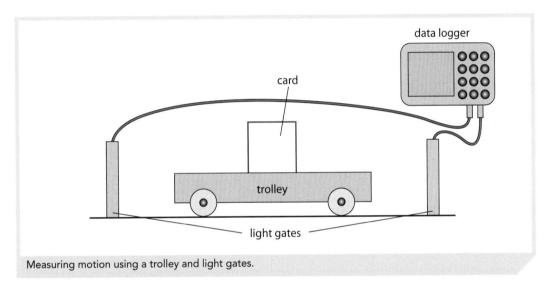

Measuring motion using a trolley and light gates.

As the trolley passes through the light gate, the card cuts the light beam.

Activity B

GPS measures distance using information from satellites. GPS is used in many applications. For example, car navigation systems use GPS to keep track of the car's position, and ships use GPS to navigate at sea.

1 Find out what GPS stands for.

2 How does a GPS satellite communicate with a GPS device on Earth?

GPS receivers can be integrated into devices such as mobile phones.

Just checking

1 Name two methods of measuring distance.

2 What is the difference between distance and displacement?

Lesson outcome

You should be able to measure distance and time in simple investigations.

6.2 Calculating speed and velocity

Key terms

Gradient – How steep a straight line is. A steeper line has a larger gradient.

Velocity – Speed in a particular direction.

If you were a top male athlete, you might run 100 m in about 10 s. Every second, you would run a tenth of the distance, 10 m. So your average speed for the race would be 10 m/s.

Distance–time graphs and speed

If you know how far something or someone has moved, and the time it took, you can work out the speed of travel. The equation that links speed, distance and time is:

$$\text{speed (m/s)} = \frac{\text{distance (m)}}{\text{time taken (s)}}$$

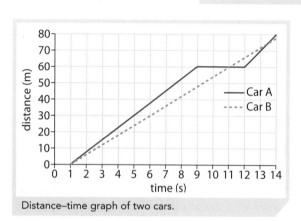

Distance–time graph of two cars.

If the distance travelled is plotted against time on a graph, the **gradient** (slope) of the graph is equal to the speed. The steeper the gradient of the graph, the greater the speed.

The graph opposite shows distance–time plots for two cars, A and B. Between 1 and 9 s the gradient of the graph for car A is greater than for car B, showing that car A is moving faster during that period. The graph also shows that between 9 s and 12 s car A does not move.

In cars, the speed is measured by sensors and displayed on a speedometer.

Speed tells you how fast a car is moving, but it doesn't tell you which direction it is moving in.

Velocity and displacement

Velocity is speed in a particular direction. For example, if you were travelling from London to Edinburgh on a train at a speed of 35 m/s, your velocity would be 35 m/s north. If you were returning from Edinburgh to London, your speed would still be 35 m/s, but your velocity would be 35 m/s south.

Velocity is given by the equation:

$$\text{velocity (m/s)} = \frac{\text{displacement (m)}}{\text{time taken (s)}}$$

Note that in the equation for velocity we use displacement, not distance, because displacement has a direction. If you plot a graph of displacement against time, the slope of the graph gives the velocity.

Remember

The gradient of a straight line can be calculated as:

$$\text{Gradient} = \frac{\text{change in } y}{\text{change in } x}$$

$$\text{gradient} = \frac{(y_2 - y_1)}{(x_2 - x_1)}.$$

In the graph above, y = distance and x = time.

Link

Look at the image in lesson 6.1 to remind yourself how light gates are used.

Worked example

In a laboratory experiment, a trolley with an 8 cm-long card passes through a single light gate. The digital timer records a time of 0.5 s. Calculate the speed of the trolley, in m/s.

Step 1 Convert cm to m: there are 100 cm in 1 m, so 8 cm = $\frac{8}{100}$ = 0.08 m

Step 2 Use the equation: speed = $\frac{\text{distance}}{\text{time}}$ and substitute for distance and time.

Step 3 Speed = $\frac{0.08 \text{ m}}{0.5 \text{ s}}$ = 0.16 m/s

This is the average speed – we are assuming that the speed does not change as the trolley passes through the light gate.

Fadil is a traffic police officer who spends most of his time investigating motor accidents. The investigations typically involve finding out how far vehicles have travelled before a collision has occurred using, for example, a measuring wheel. Sometimes, by taking measurements of oil marks on the road, Fadil can work out how fast a vehicle was travelling.

Activity A

The diagram shows a pattern of oil drops for a car involved in an accident. The dots are points on the road where oil leaked from the car engine. One drop of oil was produced every 0.5 seconds. The positions of the oil drops have been drawn to scale on the diagram; 1 cm represents 8 metres.

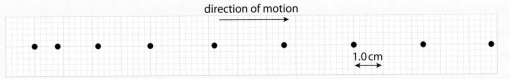

direction of motion

1.0 cm

1 What feature of the diagram shows that, finally, the car was travelling at constant speed?

2 What distance did the car travel in the last 2 seconds?

3 Estimate the final constant speed of the car.

A photo meter measuring wheel is used to measure distance following an accident.

Assessment activity 6.1

| 2A.P1 (part) | 2A.P2 | 2A.M1 (part) | 2A.D1 (part)

You have just started working as a junior technician for a car manufacturer. Your first task is to measure the speed of a car along six sections of a test track. You have measured the distance of each stretch of track, and you use sensors to measure the time the car takes to travel along each stretch. The results are shown in the table.

Section	Time (s)	Distance (m)
A	10	500
B	6	120
C	10	250
D	25	400
E	17	340
F	20	520

1 Sections A to F together make up a complete lap of the track. Plot a graph of distance against time for the whole lap.

2 Calculate the average speed of the car in mph for each section (1 m/s = 2.2 mph).

3 One section of the track has tight corners. From the results, which section do you think this is?

4 Another section has a long straight. Which section do you think this is?

5 The car had the same speed over two sections of the track.
  (a) Which were these two sections?
  (b) Would the velocity of the car also be the same in these two sections? Give a reason for your answer.

Just checking

1 What feature of a distance–time graph gives you the speed?

2 What is the difference between speed and velocity?

6.3 Acceleration

In real life, most objects change their speed or velocity as they move. This is known as non-uniform motion. When an object changes its velocity, it is accelerating. The object could be increasing in speed – this is positive **acceleration** – or its speed could be reducing – this is negative acceleration. It could also be changing direction, because velocity is speed in a particular direction. So if a car goes round a corner at constant speed, it is actually accelerating, because its velocity is changing.

The link between velocity and acceleration is given by the following equation:

$$\text{acceleration (m/s}^2\text{)} = \frac{\text{change in velocity (m/s)}}{\text{time taken (s)}}$$

$$\text{change in velocity} = \text{final velocity} - \text{initial velocity}$$

Worked example 1

1. You get on a bus at the bus stop. It pulls away. After 8 seconds it is travelling at 5 m/s. Calculate the acceleration of the bus.

   Step 1 $\text{Acceleration} = \dfrac{\text{change in velocity}}{\text{change in time}}$

   Step 2 $\text{Acceleration} = \dfrac{(\text{final velocity} - \text{initial velocity})}{(\text{final time} - \text{initial time})} = \dfrac{(5-0)}{(8-0)} = 0.63 \text{ m/s}^2.$

2. The bus driver suddenly sees an obstruction on the road and brakes abruptly. The bus slows down from 8 m/s to a complete stop in 2 seconds. Calculate the bus's acceleration.

   Step 1 $\text{Acceleration} = \dfrac{\text{change in velocity}}{\text{change in time}}$

   Step 2 $\text{Acceleration} = \dfrac{(\text{final velocity} - \text{initial velocity})}{(\text{final time} - \text{initial time})} = \dfrac{(0-8)}{(2-0)} = -4 \text{ m/s}^2.$

The acceleration of a model vehicle can be investigated by using light gates, as shown in Worked example 2. In this example, the card on the model vehicle has two sections that break the light. By measuring the length of each section and inputting this into the data logger, the timer will measure the time at four points. This information can be used to calculate an initial and a final velocity. The acceleration can then be calculated using the equation given above.

Did you know?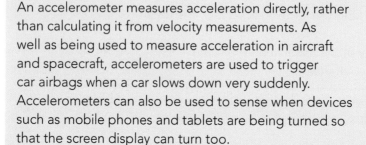

An accelerometer measures acceleration directly, rather than calculating it from velocity measurements. As well as being used to measure acceleration in aircraft and spacecraft, accelerometers are used to trigger car airbags when a car slows down very suddenly. Accelerometers can also be used to sense when devices such as mobile phones and tablets are being turned so that the screen display can turn too.

Worked example 2

A trolley with an 8-cm-long card travels along a stretch of track with two light gates on it. At light gate 1 it takes 0.6 s for the card to go through. At light gate 2 it takes 0.2 s. The trolley takes 3 s to travel from gate 1 to gate 2. Use these measurements to calculate the trolley's acceleration.

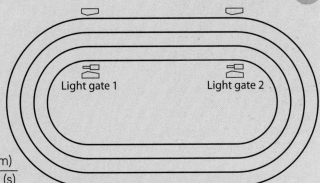

Light gate 1   Light gate 2

Step 1 As in lesson 6.2, we can calculate the speed of the trolley from the equation: $\text{speed} = \dfrac{\text{distance (m)}}{\text{time taken (s)}}$

Step 2 Speed of trolley at light gate 1 $= \dfrac{0.08\,\text{m}}{0.6\,\text{s}} = 0.13\,\text{m/s}$

Step 3 Speed of trolley at light gate 2 $= \dfrac{0.08\,\text{m}}{0.2\,\text{s}} = 0.4\,\text{m/s}$

Step 4 $\text{Acceleration} = \dfrac{\text{change in velocity}}{\text{time taken}} = \dfrac{(\text{final velocity} - \text{initial velocity})}{\text{time between gates}}$

Step 5 Substitute numbers into equation: $\dfrac{0.4\,\text{m/s} - 0.13\,\text{m/s}}{3\,\text{s}} = \dfrac{0.27}{3} = 0.09\,\text{m/s}^2$

Activity A

You are a trainee engineer working for a highways agency. You have been tasked to review the various ways that the agency is trying to catch drivers who exceed the speed limit.

1 Find out about these three techniques and describe each one:

(a) Gatso speed cameras **(b)** Truvelo speed cameras **(c)** police radar guns.

2 Which one do you think would be the most effective method for catching speeding drivers?

Just checking

1 An athlete starts from rest and accelerates to 10 m/s in 2 s. Calculate her acceleration.

2 A car travelling at 10 m/s slows down to 4 m/s over 8 s. Calculate the car's acceleration.

Lesson outcome

You should know how to use the equation for calculating acceleration.

6.4 Distance travelled

You learned in lesson 6.2 how *distance*–time graphs are useful in showing how an object changes its position with time. *Velocity*–time graphs can be used to describe the motion of objects that are changing speed. The velocity–time graph below shows how the velocity of a train varies during its journey.

(1) Between 0 and 300s the train's velocity increases at a steady rate – it has positive acceleration. On the graph this is shown as a constant gradient.

(2) Between 300 and 600s the train is at cruising speed. The velocity does not change, so the acceleration is zero. This is shown on the graph as a horizontal line.

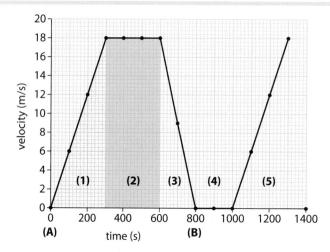

(3) As the train approaches the station, it begins to slow down; it has negative acceleration. This is shown on the graph between 600 and 800s.

(4) Between 800 and 1000s the train is in the station, and its velocity is zero.

(5) Between 1000 and 1300s the train has positive acceleration as it pulls away from Barnsley station.

Velocity–time graph for a train journey from **(A)** Sheffield to **(B)** Barnsley.

The distance travelled by the train can be obtained by calculating the area under the velocity–time graph.

Between 0 and 300 seconds **(1)**, the area under the graph is a triangle so the distance travelled in this time will be the area of the triangle. In this case the area will be $0.5 \times 18 \times 300 = 2700$ m.

The area under the graph between 300 and 600 seconds **(2)** is a rectangle and so the distance travelled will be $18 \times 300 = 5400$ m. If the total distance travelled is required then the areas of shapes **(1)**, **(2)** and **(3)** need to be added together.

If the area under the graph is not a familiar shape, such as a triangle or rectangle, then you could make an approximation of the area, for example, by counting squares.

> **Remember**
>
> Area of a triangle
> = 0.5 × height × base
>
> Area of a rectangle
> = length × width

Did you know?

A tachograph is an instrument that automatically draws speed–time graphs. Tachographs are fitted in vehicles such as coaches and lorries, which travel long distances. The tachograph provides a record to show that a driver has stayed within the speed limit, and that they have not driven for longer than is allowed by law.

Activity A

A car travels at a constant speed of 12 m/s for 30 s then slows down to 4 m/s over the next 10 s.

1 Draw a graph of the car's motion.
2 Calculate the distance the car travels over the 40 s.
3 Calculate the car's acceleration during the last 10 s.

Assessment activity 6.2   2A.P1 | 2A.M1 | 2A.D1 [part]

You are applying for a junior technician job with a company that makes model racing cars. As part of the interview you need to demonstrate your understanding of velocity and acceleration. You are given the table of data shown here, collected from experiments with a model car using motion sensors (similar to light gates).

Time (s)	Velocity (cm/s)
0.02	10.0
0.04	12.6
0.06	16.6
0.08	25.0
0.10	30.0
0.12	33.3
0.14	35.0
0.15	35.0
0.16	35.0
0.18	33.8
0.20	32.8
0.22	30.0
0.24	28.4
0.26	26.0
0.28	24.0
0.30	21.0
0.32	16.0

1 Tabulate the velocity in SI units for the total run of the model car.

2 Plot a velocity–time graph of the journey.

3 Explain the motion of the entire journey.

4 Calculate the acceleration for one part of the journey from the gradient of the graph line.

5 Estimate the total distance travelled from the area under the graph.

Tips

For 2A.P1, plot the graph points accurately and choose your graph scales carefully. The assessor is looking for accurate plotting and for the data to fill most of your graph paper.

For 2A.M1 make sure you use scientific words in your explanation.

For 2A.D1 remember that acceleration and distance have units.

Lesson outcomes

You should be able to produce graphs that represent uniform and non-uniform motion, and be able to interpret them.

Just checking

1 What information can you get from a velocity-time graph?
2 What does it mean if an object has negative acceleration?

6.5 Energy transfers

When you go on a rollercoaster, the first part is very slow. A chain drive hauls the carriages up to the top of the first rise. Then you suddenly plunge downwards, making you feel like you've left your stomach far behind. The track swoops upwards and down again, and on some rides you loop-the-loop and turn corkscrews. But after the first slope, it is all unpowered – so how does it work?

Gravitational potential energy

Rollercoaster rides rely on **gravitational potential energy** (GPE). A carriage at the top of the first rise of a rollercoaster ride has stored GPE.

The amount of stored GPE depends on how high the rollercoaster is. This can be calculated using the equation:

> GPE (J) = mass (kg) × acceleration due to gravity (m/s^2) × change in height (m)
>
> $$\text{GPE} = m \times g \times h$$

Worked example 1

Calculate the gravitational potential energy for the rollercoaster carriage at A in the diagram. Take the acceleration due to gravity as $10\,\text{m/s}^2$. The mass of the carriage is 150 kg.

Step 1 Write down the equation:
GPE = $m \times g \times h$

Step 2 Substitute values into the equation:
GPE = 150 kg × 10 m/s^2 × 200 m = 300 000 J

Remember to include the units in your calculations.

From GPE to KE

The law of conservation of energy states that energy cannot be created or destroyed, it simply changes from one form into another. As a rollercoaster carriage starts going downwards from the top of a track, its height decreases and so the GPE decreases. However, at the same time the speed of the carriage increases, so the GPE does not simply disappear. The gravitational potential energy is transformed into **kinetic energy** (KE).

The KE of an object depends on its mass and its speed. The faster an object is moving and the greater its mass, the greater its KE. KE can be calculated using the equation:

> kinetic energy = 0.5 × mass × (speed)^2
>
> $$\text{KE} = \tfrac{1}{2} \times m \times v^2$$

Notice that KE is proportional to speed *squared*, not simply to speed. The worked example shows what effect this has.

Worked example 2

A car carrying two passengers has a total mass of 800 kg. Calculate its kinetic energy at a speed of 12 m/s.

The car doubles its speed. What is its kinetic energy at the new speed?

Step 1 Write down the equation: KE = ½ × m × v^2

Step 2 Substitute values into the equation for a speed of 12 m/s:
KE = 0.5 × 800 kg × (12 m/s × 12 m/s) = 400 × 144 = 57 600 J.

Step 3 Do the same calculation for double the speed:
KE = 0.5 × 800 kg × (24 m/s × 24 m/s) = 400 × 576 = 230 400 J.

So at double the speed, the car has four times (2^2 = 2 × 2) as much kinetic energy.

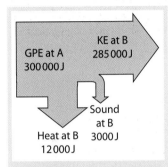

Energy transformation diagram showing how energy is transformed as a rollercoaster carriage travels from A to B. Not all the GPE is transformed to KE. Some is transformed into heat and some becomes sound.

Energy transformation diagrams

At B in the rollercoaster diagram on the left, the GPE of the carriage is at its lowest. Most of the GPE has been converted to KE, but **friction** between the wheels and the track means that some energy is transformed into heat and sound. This can be represented by an **energy transformation diagram**, as shown on the right.

As the rollercoaster carriage climbs from the bottom of the first dip (B) to the top of the second rise (C), it gradually slows down. Some of its KE is transformed back to GPE. As it continues along the track, the energy of the carriage keeps shifting between KE and GPE. The useful energy that the carriage has gradually reduces, as some energy is lost as heat and sound. If this did not happen, the rollercoaster could keep going for ever!

Stopping distance

Kinetic energy helps us calculate the distance a car will travel after the brakes are applied. This is called the **braking distance**. When the brakes are applied, the brake pads transfer the car's kinetic energy into heat.

The time it takes for the driver to react to a hazard before applying the brakes is called the **thinking distance**.

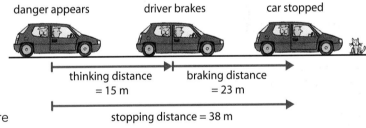

danger appears   driver brakes   car stopped

thinking distance = 15 m   braking distance = 23 m

stopping distance = 38 m

total **stopping distance** = thinking distance + braking distance

Activity B

Which distance, *thinking* distance or *braking* distance, is calculated from the equation KE = ½ × m × v^2? Give a reason for your answer.

Just checking

1 It is 35 m from the top of a rollercoaster track to the bottom. What is the GPE of a rollercoaster carriage at the top, with a mass of 500 kg and two passengers with masses of 50 kg and 75 kg? Use a value of 10 m/s² for the acceleration due to gravity.

2 What is the KE of the carriage at the bottom of the track, assuming that 5% of the total energy is lost as heat and sound?

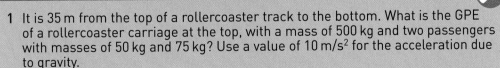

Lesson outcomes

You should be able to use the equation for gravitational potential energy and explain how it is transferred to kinetic energy; describe energy transfers using energy transformation diagrams; and explain stopping distances.

6.6 Investigating conservation of energy

Get started

Look at the apparatus in the diagram opposite. Discuss with a partner what you think would happen to the marble on this track. What energy transformations would occur? What will happen eventually?

Key term

Conservation of energy – The principle that energy cannot be created or destroyed, only transformed from one form to another.

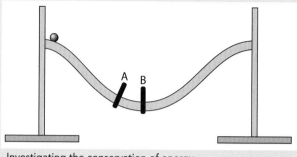

Investigating the conservation of energy.

The law of **conservation of energy** states that energy cannot be created or destroyed: it can be changed into different forms, for example, GPE to KE, or transferred from one object to another, but the total energy will always remain the same.

You can model a rollercoaster using a marble on a track as shown in the diagram. A and B are light gates connected to a data logger. You can use this apparatus to investigate the conservation of energy.

Measure the mass of the marble using a digital balance. Roll the marble down the track starting from different heights. You can use light gates A and B to work out the speed of the marble at the bottom of the track. The data logger measures the time it takes the marble to travel from A to B.

From the mass of the marble and the speed measurement you can calculate the gravitational potential energy of the marble at the starting point and its kinetic energy at the bottom of the curve, and compare them.

What energy changes will occur during this rollercoaster ride?

Worked example

In an investigation using the apparatus shown above, the marble has a mass of 10 g and the distance between A and B is 10 cm. The investigation produces the following data:
starting height = 0.06 m
time between A and B = 0.096 s.

1. Calculate the gravitational potential energy.
   Step 1 Use the equation: GPE (J) =
   mass (kg) × acceleration due to gravity (m/s^2) × change in height (m)
   Step 2 Convert the mass of the marble into kilograms: 10 g = 0.01 kg
   Step 3 Height 0.06 m: GPE = 0.01 kg × 10 m/s^2 × 0.06 m = 0.006 J

2. What is the predicted kinetic energy according to the conservation of energy, assuming no loss of energy as heat or sound?

The predicted KE is the same as the GPE: 0.006 J.

3. Calculate the average speed based on the time measured by the data logger.
   Step 1 Use the equation: speed (m/s) = distance (m) ÷ time (s)
   Step 2 Convert the distance between A and B to metres: 10 cm = 0.1 m
   Step 3 Height: 0.06 m: speed = 0.1 m ÷ 0.096 s = 1.0 m/s

4. Calculate the kinetic energy based on the measured speed.
   Step 1 Use the equation: KE = 0.5 × mass × (velocity)^2
   Step 2 Remember to use the mass in kg.
   Step 3 Height = 0.06 m: KE = 0.5 × 0.01 kg × 1.0^2 = 0.005 J

Link

Refer back to lesson 6.5 to remind yourself of how to calculate GPE and KE; and to lesson 6.2 to calculate speed.

5 Suggest reasons for the differences between the KE values calculated in questions **2** and **4**.

The KE calculated in **Q4** is less than that predicted in **Q2**.

KE(calculated) as a percentage of KE(predicted) = (0.005 ÷ 0.006) × 100 = 83%.

The difference is possibly due to the transformation of some GPE into heat and sound.

6 Draw an energy transformation diagram for this experiment.

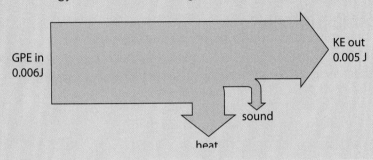

GPE in
0.006J

KE out
0.005 J

sound

heat

Activity A

Using the same apparatus and method as in the Worked example:

1 Calculate the GPE for marbles with starting heights:

(a) 0.12 m; (b) 0.24 m; (c) 0.5 m.

2 The time taken to get from A to B for these marbles was:

(a) 0.068 s; (b) 0.048 s; (c) 0.033 s.

Calculate the average speed and the KE for each marble.

Assessment activity 6.3 | 2A.P3 | 2A.M2 | 2A.D2

1 You have started a job with a funfair engineering company. As part of the training process, they want you to carry out some tests on a model rollercoaster system.

(a) Describe how you could use a rollercoaster track made from cardboard, some marbles and light gates to investigate the conservation of energy. How would you obtain data to calculate the speed of the marbles?

(b) Draw an energy transformation diagram for one part of your track.

(c) Using the energy transformation diagram you have drawn, explain how energy is being conserved in that part of the track.

(d) Your marbles weigh 25 g each.

  (i) At the bottom of one dip you record an average speed of 10 m/s. Calculate the KE of the marbles at this bottom point.

  (ii) The top point of your track is 2 m high. What is the GPE of a marble at this top point?

2 Imagine you are driving to work and a rabbit runs out in front of your car.

(a) Explain the energy transformations that will take place when you brake suddenly.

(b) What factors will affect the likelihood of you stopping before you hit the rabbit?

 Just checking

1 Explain what an energy transformation diagram tells you about a system.

2 How may energy be lost in a rollercoaster?

 Tips

When you are designing your investigation, think carefully about how it would actually work. Draw a diagram to show how the apparatus would be arranged.

For 2A.M2, it is vital that you show your working. Include the correct units with your answer.

For 2A.D2, you need to use the correct scientific terminology and include any energy that is lost in the transformations.

 Lesson outcomes

You should be able to draw and correctly label energy transformation diagrams, and be able to explain the energy changes affecting transportation and stopping distances.

6.7 Introducing forces

Get started

What is a force? In groups of three, write down a definition. Discuss how you might show forces on a diagram.

Key terms

Balanced forces – When forces are balanced, the resultant force is zero and the object remains at a constant speed.

Unbalanced forces – When forces are unbalanced, there is a resultant force and the object will change speed.

You may be amazed to know that everything you do involves some sort of force; for example, you would not be able to walk down the street without your shoes exerting a force on the pavement and the pavement acting on your shoes. A force is a pull or push that is exerted by one or more objects on another. The unit of force is the **newton** (N).

Like velocity measurements, forces have both magnitude (size) and direction. One way that you can show this is to draw forces as arrows on a diagram. The length of the arrow shows the size of the force, and the point of the arrow shows its direction.

Action–reaction pairs

Forces are interactions between two objects. They come in pairs. For example, suppose you push against a wall as in diagram **(a)**. You exert a forwards force on the wall, but you can also feel the force of the wall pushing back against you. This pair of forces is called an **action–reaction pair**. Action–reaction pairs are equal and opposite; they are the same type of force, but act on different bodies.

Diagram **(b)** shows another action–reaction pair. The backwards force on the starting block from the sprinter's foot and the forwards push of the block on her foot. Although the action–reaction forces are equal and opposite, the effects on the sprinter and on the starting block are very different. The backwards force on the starting block from the sprinter's foot is balanced by an opposite force: the push of the ground against the block. So the block does not move. However, there is no opposite force on the sprinter to balance the forwards force from the starting block. The result is that she surges forward along the track.

Two examples of action–reaction pairs. In example **(a)** there is no movement, although the wall is very slightly compressed, but in **(b)** the sprinter powers forwards.

Balanced and unbalanced forces

Forces can be either **balanced** or **unbalanced**. For example, an apple on a desk is stationary because the forces on it are balanced. The force of gravity trying to pull the apple downwards, the apple's **weight**, is balanced by the force of the desk pushing the apple up. The **resultant force** on the apple is zero.

If an object is moving at constant velocity, the forces on it are also balanced. Suppose a car is driving along a straight road in a town at 30 mph. The car engine provides a forwards force, but this is balanced by two backwards forces: friction between the car's tyres and the road; and **air resistance** or **drag**. The resultant force is zero, even though the car is moving.

Activity A

Look at the ice skaters below. When they push against each other what will happen? Include a force diagram and ideas about action–reaction forces and unbalanced forces in your answer.

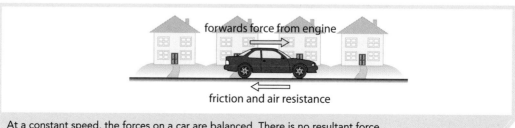
forwards force from engine

friction and air resistance

At a constant speed, the forces on a car are balanced. There is no resultant force.

When the driver reaches the edge of town and the speed limit changes, she presses on the accelerator and the engine produces more power. Now the forces on the car are unbalanced. The forwards force is greater than the backwards forces from friction and drag. There is a forwards resultant force, and the car accelerates forwards.

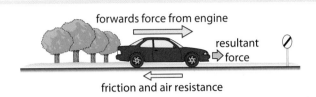

forwards force from engine

resultant force

friction and air resistance

When the driver accelerates, the engine produces more force, and the forces are unbalanced. The resultant force is forwards.

Case study

Peter is a health and safety inspector. One of his jobs is to check safety on building sites. Peter checks that cranes are loaded correctly and that concrete slabs are lifted safely. He also looks at the way that walls are supported while they are being built. He makes sure they are braced by hydraulic pistons running diagonally from the wall down to a trench in front of the wall. Peter also looks around the site to make sure that there are clear signs to identify hazards.

1 Describe how many of the things that Peter inspects involve forces.

2 Describe the action–reaction forces involved in bracing the walls.

Assessment activity 6.4 | 2B.P4 | 2B.D3

You are a junior mechanical engineer preparing a presentation on forces to use at an open day for schools.

1 Draw force diagrams that show the size and direction of forces on
   (a) a book resting on a table
   (b) a car moving at constant speed.

2 Draw force diagrams that show the size and direction of forces on
   (a) a train leaving a station
   (b) an aircraft slowing down on the runway by applying wheel braking.

3 Using the answers to questions **1** and **2**, explain the forces on each object.

Just checking

1 What is meant by balanced forces?
2 What are action–reaction pairs?
3 What happens to an object with a resultant force?

Tips

What is really important is that the size and direction of the arrows are clearly shown and are correct.

For 2B.P4 you need to explain the size and direction of the forces, and what is happening to the book, car, train and aircraft, using scientific terminology.

For 2B.D3, the key is to explain using diagrams; this will make your explanations clearer.

Lesson outcomes

You should be able to explain how forces arise, understand action–reaction pairs, and use your knowledge of balanced and unbalanced forces to draw and explain force diagrams for objects at rest and in motion.

6.8 Forces in action

Get started

When you go down a long hill on a bike, you get faster at first but then your speed reaches a maximum. Why do you think this happens? Discuss possible reasons in pairs.

When an engineer designs a new car, aircraft or boat, a very important thing to consider is drag. When an object moves through air or water it experiences a **drag force** that slows it down. This drag force is sometimes called air resistance or water resistance. Reducing drag can improve performance and reduce fuel costs.

A rocket flight

The diagram below shows the stages of a rocket flight. How do drag forces interact with other forces during the flight?

- As the rocket blasts off, it has to push through the air particles surrounding it.
- This causes a drag force (air resistance) which acts against the upward motion of the rocket. As the rocket moves faster, the air resistance (drag) gets greater.
- At a certain height the rocket cuts its motors. It is still travelling upwards, but the only forces acting on the rocket are its weight (downwards) and the drag (also downwards).
- The resultant force is opposite to the direction of motion, so the rocket begins to slow down.
- Eventually the downwards forces slow the rocket's speed to zero. Now it is at its maximum altitude (height).

The smoke trails show the air flow over the sports car. The flattened shape of racing cars minimises their air resistance.

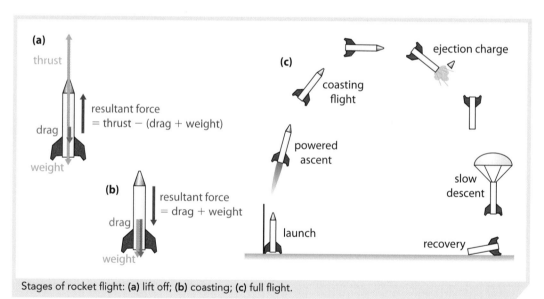

Stages of rocket flight: **(a)** lift off; **(b)** coasting; **(c)** full flight.

- At zero velocity, the only force on the rocket is its weight (the force of gravity acting on the rocket's mass). This is downwards, so the rocket begins to accelerate downwards.
- The downward movement of the rocket produces air resistance, but this time it acts *upwards*. At first this drag force is small, but as the rocket falls faster, the drag increases.

If the rocket continued to fall unchecked, it would go faster and faster until the downwards force of its weight was balanced by the upwards force from air resistance. At this point the rocket would stop accelerating and fall at a constant speed. This is called **terminal velocity**.

In fact, the rocket does not continue to fall – its parachute opens. The parachute has a large surface area, and this greatly increases the drag forces on the rocket. The upward drag forces are now greater than the rocket's weight, and it begins to slow down.

As the rocket slows, the air resistance gets smaller. At a certain velocity, the air resistance reaches a level where it exactly balances the weight. The rocket is now falling at terminal velocity, but this terminal velocity is much slower than it would have been without the parachute.

What forces will these skydivers experience during their descent?

Activity A

1 Use the example of the rocket to help you to describe and explain the forces on a skydiver in free fall, and after their parachute has opened.

2 Research the terminal velocity reached by skydivers in free fall, and after they open their parachutes.

3 The diagram shows the forces acting on a car towing a caravan at a constant speed of 50 mph.

D = the driving force of the engine
F = the frictional and drag forces on the car
f = the frictional and drag forces on the caravan
W = the weight of the car
w = the weight of the caravan

$R + r$

$F + f$ ← T D →

$W + w$

(a) What is the force ($R + r$)? Note that this force has the same value as ($W + w$) but in the opposite direction.

(b) If D = 3100 N and f = 1200 N, calculate the frictional and drag forces F acting on the car.

(c) Why does the car have a maximum speed? Use ideas about terminal velocity in your answer.

(d) Why is the car's top speed likely to be greater when it is not towing a caravan?

Just checking

1 What is another word for air resistance?
2 How can air resistance be minimised?
3 Can you think of situations where you might want to increase air resistance?

Lesson outcomes

You should be able to explain the forces acting on a rocket during flight, on a parachutist and on a car during braking and acceleration.

6.9 Friction forces

Get started

In pairs, discuss what you have already learned about friction in this unit.

Key terms

Applied force – A force that is applied to an object by a person or by another object.

Friction – The force between two surfaces that resists motion.

Normal force – A force acting at 90° to a surface.

Sometimes the force of friction is a real nuisance. Friction can quickly cause an engine to overheat and seize up, so engines need constant cooling and **lubrication** to keep them working. More often, friction is incredibly useful. Without friction your bike wheels would skid when you turned the pedals, you would not be able to walk, and anything you tried to pick up would simply slip through your fingers.

What is friction?

Friction forces occur in response to an **applied force** between materials that slide or rub against each other. Friction forces act to stop an object from moving, or to slow it down when it is in motion.

The diagram shows a crate being pushed to the right (red arrow). This produces a friction force between the crate and the floor (green arrow) that opposes the applied force (the push). There

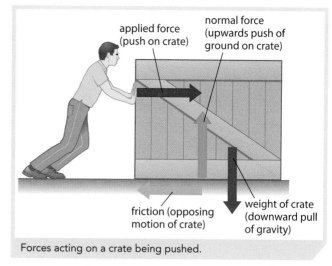

Forces acting on a crate being pushed.

are also other forces acting on the crate: its weight; and the **normal force**. The normal force acts at 90°; it is the upwards push of the floor on the crate.

If the crate is only pushed gently, it will not move, because the push is balanced by friction. If the push increases a little, friction increases too. However, at a certain point the push becomes big enough to overcome friction and the crate will start to move. The size of push needed to get the crate moving depends on the weight of the crate and on the surfaces of the crate and the floor.

Investigating friction

Material	Force (N)
Sandpaper	5
Carpet	4
Work surface	3

The diagram below shows how you could investigate which material produces the greatest friction when an object is dragged across it. In this case, a wooden block is tested on a piece of carpet, some sandpaper and a smooth wooden work surface. The force measured by the force meter just as the block starts to slip is called the **static frictional force**. The table on the left shows the kinds of results you might get from the investigation.

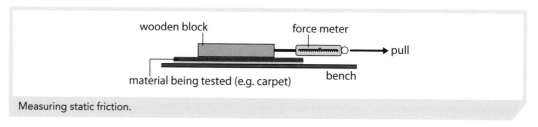

Measuring static friction.

Remember

Weight is a force, measured in newtons.

You can use the same apparatus to investigate the effect that changing the weight of the wooden block has on the static frictional force. You will find that weight has a larger effect than the type of surface. For example if you double the weight, you will need to double the pulling force.

Friction and heat

Friction between two objects or materials can cause them to heat up. For example, if you brake hard on your bike, friction between the brake blocks and the wheel will heat up the brakes. The kinetic energy of the bike is transformed into thermal energy.

Friction in the brake discs of a Formula 1 racing car is much greater than in bicycle brakes. The brake discs become red-hot during a race. Designing systems to remove the thermal energy is an important part of the work of a braking system engineer.

Friction helps vehicles to accelerate forwards as well as being important for braking. Without friction between the tyres and the road, the wheels would simply skid when they turned. A smooth tyre has maximum contact with the road and therefore provides maximum road-holding in the dry. However, in wet conditions water gets between the tyre and the road and reduces friction. To avoid this problem, all road tyres have a tread pattern that disperses water.

Ice also affects friction between tyres and the road. The tyres lose their grip at lower speeds on icy roads than on normal roads, making cornering and braking more hazardous.

Friction makes the brake discs of this racing car glow red-hot.

Activity A

1 Explain why it is vital that there is a friction force that opposes motion when two surfaces try to slide across one another.
2 Give two examples of things that would be impossible to do without friction.

Assessment activity 6.5
| 2B.P6 | 2B.M4

1 In the form of a table, list THREE situations involving friction.
2 Describe an experiment that can be carried out to investigate the effect of weight on friction.
3 Explain the frictional forces involved when pushing against a box on **(a)** a smooth surface and **(b)** a rough surface.
4 Explain how the normal force is produced in a stationary car.

Tips

You need at least three examples of where friction is used; less than this number and you won't meet the criterion.

For 2B.P6, you need to link the materials (rough and smooth surfaces) and the frictional force required to make something slip. You also need to describe how the normal force is produced in response to an applied force.

Just checking

Use the idea of friction to explain how you should change the way you drive if conditions are icy. Refer to engine force in your answer.

Lesson outcomes

You should be able to identify friction forces and explain situations where they occur.

6.10 Compressive and tensile forces

Gold is often used to make jewellery because it is relatively easy to stretch and form into different shapes. However, other materials such as glass break quickly when stretched, so manufacturers of windows have to take care not to break the glass when they are fastening the panes.

Forces that cause materials to stretch are called **tensile** (pulling) forces. The opposite of tensile forces are **compressive** (pushing) forces. These are forces that cause materials to be compressed (squashed).

Elastic or inelastic

Some materials are **elastic**. They **deform** (change shape) when they are stretched or compressed, but then spring back to their original shape when the stretching or compressing force is removed. Rubber is a good example of an elastic material.

Elastic materials are used:

- in tyres, which make the ride more comfortable and reduce damage to the wheels
- as seals to prevent liquids or gases from leaking, and to reduce vibrations in engines
- in trainers to make running more comfortable
- to put the bounce in balls
- to make railways run more quietly.

Inelastic materials can also be affected by compressive or tensile forces. When inelastic materials change shape after being stretched or compressed, the change is permanent – they do not go back to their original shape when the force is removed. Metals and glass are examples of inelastic materials. Most metals are **malleable**, which means they can be shaped by hammering or pressing. Hot steel, for example, is shaped into sheets by pressing it between large rollers.

Some metals are also **ductile**, which means they can be stretched. For example, steel and copper are ductile metals. Hot copper or steel rods can be drawn out by tensile forces into wires, with uses ranging from bridge cables to electrical wiring to jewellery.

The weight of a bungee jumper acts as a tensile force on the bungee. The bungee is made of elastic material, so it stretches.

Activity A

Plan an experiment using the apparatus shown to investigate tensile forces. Start with the elastic under a tensile force of 1 N, and measure its length. Check your plan with your teacher before you begin your practical work.

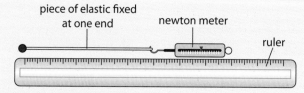

1 What do you think will happen if you increase the tensile force?

2 Measure the length of the elastic at six different values of force and plot the results on a graph.

3 Describe the relationship between length of the elastic and tensile force.

Forces on structures

A material can be under tension or compression but not appear to change shape at all.

If you hang from an overhead bar, your arms do not lengthen, but you can feel the tensile force as your weight pulls you down.

The walls or pillars of a building do not squash, but they are under compression from the weight of the floors, roof and other structures they are supporting.

A beam between two supports is under both tension and compression. If the beam was made of a material that did bend, then if you put a weight in the centre, it would bow downwards. The bottom of the beam would stretch and the top would be compressed. Most beams do not bend under a load, but the top of the beam is still under compression while the bottom is under tension.

The Tsing Ma Bridge in Hong Kong is one of the longest suspension bridges in the world. The towers are made from concrete to withstand the compressive weight of the cables. The cables themselves are made of steel, which can withstand strong tensile forces.

Understanding tensile and compressive forces is important for all kinds of construction work. In a suspension bridge the cables that support the bridge deck are under tension, while the pillars that hold up the suspension cables are mainly under compression. Engineers must understand these forces in order to choose the right material for each part of the bridge. Steel is used for the cables because it is good at resisting tensile forces. Concrete is strong under compressive forces, so it is often used for bridge pillars. The concrete is often reinforced with steel to make it even stronger.

Did you know?

The London Millennium Footbridge is a steel suspension bridge that opened in June 2000. Because of its unusual design, the suspension cables carry an unusually high tensile force. When people first started walking on the bridge on the day it opened, they noticed that it swayed significantly, so much that some people had to grab hold of the handrail to steady themselves!

The sway was considered to be so alarming and potentially dangerous that the bridge was closed 2 days later for investigation and repair and didn't reopen until 2 years later. Even today, some people still call the London Millennium Footbridge the 'wobbly bridge'.

The London Millennium Footbridge.

Just checking

1 Explain what is meant by **(a)** a compressive force, and **(b)** a tensile force.
2 Describe how **(a)** elastic and **(b)** inelastic materials react under these forces.

Lesson outcome

You should know some applications of compressive and tensile forces.

6.11 Force and acceleration

Get started

How do seat belts and air bags help to protect you in a car crash? Think about the forces involved in a collision.

Link

In lesson 6.3 you learned that acceleration is equal to change in velocity divided by the time taken.

Table 1

Force (N)	Mass (kg)	Acceleration (m/s²)
1	1	1
2	1	2
3	1	3

Table 2

Force (N)	Mass (kg)	Acceleration (m/s²)
1	1	1
1	2	0.5
1	4	0.25

Cars speed up, slow down or travel at constant speed because of the relationship between the engine force and the drag forces acting on the cars.

The diagram below shows an investigation into the relationship between force, mass and acceleration. The apparatus uses a trolley, light gates and a pulley system. Weights attached to the trolley by a string provide the pulling force. The card on the trolley and the light gates measure the acceleration between points A and B.

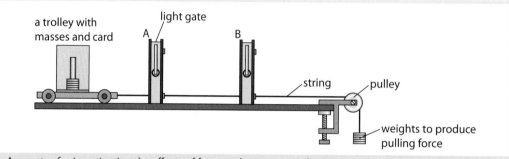

Apparatus for investigating the effects of force and mass on acceleration.

In the first part of the investigation you measure the effect of varying the force. Keep the total mass of the system constant by starting with all the weights stacked on the trolley, then varying the pulling force by transferring different numbers of weights to the end of the string. Table 1 shows some typical results.

In the second part of the investigation the force is constant at 1 N but the mass of the trolley varies. Table 2 shows some typical results.

The investigation shows that the acceleration of an object depends on the size of the force applied to it and on the object's mass. If the force doubles then the acceleration will double. If the mass doubles then the acceleration will be halved.

This relationship is described by the following equation:

$$\text{force (N)} = \text{mass (kg)} \times \text{acceleration (m/s}^2)$$
$$F = m \times a$$

This can be rearranged as:

$$\text{acceleration (m/s}^2) = \text{force (N)} \div \text{mass (kg)}$$
$$a = \frac{F}{m}$$

Worked example

The mass of a car and its two passengers is 800 kg. The engine force is 3000 N, air resistance is 800 N and frictional force is 1000 N.

1 Calculate the acceleration.

    Step 1 Calculate the resultant force:

        resultant force = engine force − air resistance − frictional force

                = 3000 N − 800 N − 1000 N = 1200 N

    Step 2 Rearrange the equation $F = m \times a$ to make acceleration the subject:

$$a = \frac{F}{m}$$

    Step 3 Substitute numbers into the equation: $a = 1200\,\text{N} \div 800\,\text{kg} = 1.5\ \text{m/s}^2$

2 What would the acceleration be if the car picked up another two passengers, with combined mass 200 kg?

Step 1 Resultant force stays the same = 1200 N.

Mass has increased to 800 kg + 200 kg = 1000 kg.

Step 2 $a = \dfrac{F}{m}$ = 1200 N ÷ 1000 kg = 1.2 m/s^2

Note that the acceleration has decreased. This would be expected, as the mass is greater but the force is unchanged.

FO4305OZ02

Seat belts and air bags help to protect you in a car crash.

Activity A

You are a motor sport technician designing a braking system. Your first task is to find out what braking force is required to stop a racing car that is travelling at 50 m/s. The mass of the racing car plus driver is 650 kg and you need the car to stop in 5 s.

Calculate the braking force required.

Hint: First you need to find the negative acceleration, using information from lesson 6.3.

During a Formula 1 race, engineers receive all kinds of information from sensors on the car. One kind of sensor measures the car's acceleration second by second.

Assessment activity 6.6 | 2B.M3 | 2B.D3 |

1 A rocket has a mass of 5 kg and lifts off with an acceleration of 12 m/s^2. Calculate the rocket's thrust if the drag is 500 N.

2 Explain the forces acting in the following applications:

   (a) a rocket from launch through engine cut-off to reaching zero upward velocity

   (b) a car that accelerates from rest, reaches a constant speed, and then slows when the brakes are applied until it comes to rest again.

Tips

For 2B.M3, the answer must have units and your working out must be clearly shown.

For 2B.D3, the best way to explain the forces is to use diagrams; otherwise your description may not be clear and so will not meet the criterion.

Just checking

1 What happens to the acceleration when an object's mass is halved, assuming the force doesn't change?

2 What happens to the acceleration when the force is doubled and the mass is doubled?

Lesson outcome

You should be able to use the equation relating force to mass and acceleration.

Key terms

Work done – A type of mechanical energy, explained by the equation:

work done = force × distance moved in the direction of the force.

Activity A

Calculate the braking distance of a car that has a resultant force of 3 kN acting on it and uses 30 kJ of work to come to a complete stop.

Remember that 1 kN = 1000 N and 1 kJ = 1000 J.

Whenever you do any kind of activity, your body has to do work. The work could be as simple as lifting a coffee cup or going upstairs. Or it could involve doing a high-energy dance routine or hauling a sledge on an Arctic expedition. All these examples involve a force moving something.

What is work done?

Work done is the energy needed to move something. The work done depends on:

- the amount of force needed to move the object
- the distance that it is moved in the direction of the force.

The equation used to describe the work done is:

$$\text{work done (J)} = \text{force (N)} \times \text{distance (m)}$$

Work done is a measure of energy, so the units are joules (J). Work done is the energy transformed when a force moves through a distance and makes an object move.

Worked example 1

Calculate the work done when you push a shopping trolley 200 cm. The force applied is 100 N.

100 N

200 cm

Step 1 Write down the equation:
work done (J) = force (N) × distance (m)

Step 2 The distance is given in centimetres (cm).
Change this to metres by dividing by 100.
distance = 200 cm ÷ 100 = 2 m

Step 3 Substitute the values of force and distance into the equation:
work done = 100 × 2 = 200 J

Investigating work done

Worked example 2 shows how you can compare the work done when moving a box horizontally with the work done when moving it vertically. A **newton meter** is a device used to measure force.

Worked example 2

Sara is a trainee engineer. She is investigating how much work is needed to **(a)** pull a box 30 cm compared to the work needed to **(b)** lift it 30 cm. She uses a newton meter to measure the force. The diagrams below show the experiments.

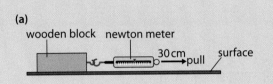

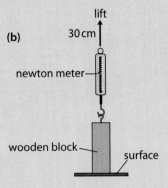

The results are shown in the table.

	Force (N)	Distance (m)	Work done (J)
Lifting	16	0.3	
Pulling	8	0.3	

Calculate the work done in both lifting and pulling. What do the results show? Explain your results.

Step 1 Work done = force × distance

Step 2 Work done lifting = 16 N × 0.3 m = 4.8 J

Step 3 Work done pulling = 8 N × 0.3 m = 2.4 J

The results show that the work done to pull a block 30 cm horizontally is smaller than the work done to lift it the same distance.

When lifting the box, the forces acting are the weight (downwards) and the applied force (upwards). In order to lift the box you need to overcome the downward forces. In the pulling experiment, the only force that you need to overcome is the frictional force between the box and the workbench.

Assessment activity 6.7 | 2B.P5

You are a mechanical engineer investigating how to reduce the forces needed when transforming energy.
1 Use force diagrams to describe the work done in the following applications:
   (a) pushing a supermarket trolley
   (b) pushing a box along a flat surface.

2 Carry out two experiments similar to ones in this lesson, and then use the correct equation to calculate the work done.

Just checking

100 J of work is done to push a shopping trolley 4 m. How much force is applied?

The crane has to do a lot of work to lift the giant ship propeller.

 Tips

In the force diagrams, make sure that the forces have arrows.

For 2B.P5, you need to keep a clear log of what you did and the results that you achieved. Your supervisor should sign this log to confirm that you carried out the experiments. Note also that to meet 2B.P5 you need to carry out a calculation of work done related to the experiments.

 Lesson outcomes

You should be able to describe what work done is, and know how to calculate it using the equation relating force and distance.

6.13 Investigating light

Get started

Did you look at your face in the mirror this morning? Where do you think your image is located in the mirror? Can you think of any other material that also gives you a reflection?

Key terms

Concave – Thinner in the middle than at the rim.

Convex – Thicker in the middle than at the rim.

Plane mirror – A mirror with a flat surface.

Reflected – Waves are reflected when they bounce off a surface, for example light bouncing off a mirror.

Did you know?

It takes light 8 minutes to get from the Sun to the Earth.

These laser beams show light travelling in straight lines.

Light travels in straight lines

If you have ever watched a laser light show, you will have seen that light travels in straight lines. This can be demonstrated using the equipment shown in the diagram below. We represent light travelling using **ray diagrams**. The rays of light are always drawn as straight lines.

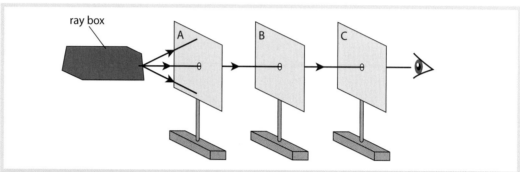

Three pieces of card (A, B and C), each with a hole in the centre, are placed in line. The ray box is positioned so you can see the light from beyond card C. If you move card C a little, you will not be able to see the light. This proves that light travels in straight lines.

Laws of reflection

Light can be **reflected** by many different surfaces. **Mirrors** have very smooth surfaces that reflect light well. You can use ray diagrams to show how light is reflected by mirrors.

The light ray that strikes the surface of the mirror is called the **incident ray** and the light ray that is reflected (leaves the surface) is called the **reflected ray**. The angle at which light strikes the mirror is measured to an imaginary line that is drawn at 90° to the reflecting surface, as shown in the diagram below. This imaginary line is called the **normal**.

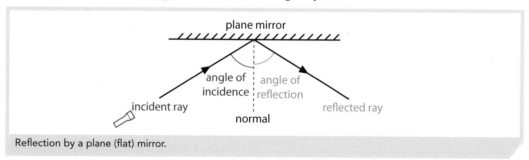

Reflection by a plane (flat) mirror.

If you carried out the experiment shown in the diagram, you would discover this law:

angle of incidence = angle of reflection

Activity A

1 List the equipment you will need to carry out an investigation that will verify (check) the law that says that the angle of incidence is equal to the angle of reflection.

2 Make a plan to carry out the investigation.

3 Ask your teacher to check your plan, then carry out the investigation.

4 Is the law of reflection correct?

Images in plane mirrors

In a **plane** (flat) **mirror**, the image is as far behind the mirror as the object is in front. If you go behind the mirror to look at the image you cannot see it. The image only *appears* to be behind the mirror and is called a **virtual image**. This is because the rays of light are reflected to our eyes as though they came from behind the mirror. Virtual images cannot be displayed on a screen.

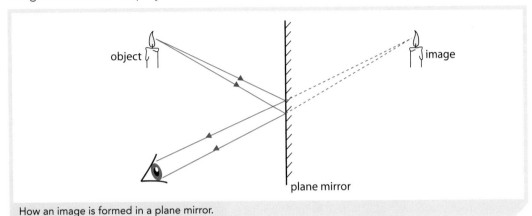

How an image is formed in a plane mirror.

Activity B

Look at your reflection in both sides of a spoon. Which side acts like a convex mirror?

Images in curved mirrors

Curved mirrors come in two kinds: **concave** and **convex**. Convex mirrors are mirrors that are curved outwards. If you look into a convex mirror, the middle of the mirror is closer to you than the edge. Concave mirrors curve inwards.

The more curved a mirror is, the more powerful it is.

Concave mirrors produce images that are magnified compared to the object.

Convex mirrors produce images that are diminished (smaller) compared to the object. A convex mirror has a wider field of view than a plane mirror.

Convex mirrors are often used in supermarkets so that staff can check that no-one is shoplifting.

Mirrors and driving

Cars have rear view mirrors so the driver can see behind without having to turn around. These mirrors are plane mirrors. Most cars also have door mirrors, to help the driver to look behind. Sometimes convex mirrors are used because the field of view is bigger. However, the image is smaller than the actual size and this makes the object appear further away than it really is. Drivers must be careful not to misjudge when it is clear to pull out.

Mirrors are sometimes used at road junctions to improve drivers' sight lines, for example, to help them see around a sharp bend. These mirrors are usually convex.

Just checking

1 What is the normal?
2 What are the angle of incidence and the angle of reflection?
3 What is the law of reflection?
4 Name two types of curved mirrors.

Lesson outcomes

You should know how to draw light rays to represent light moving in straight lines, and the laws of reflection for plane mirrors.

6.14 Mirrors in action

Get started

In groups, discuss how many applications you have at home that involve mirrors. Can you think of any applications of mirrors outside the home? Discuss how the mirrors are used.

Key term

Periscope – An instrument for observation from a concealed position. Its simplest form consists of an outer case with mirrors at each end set parallel to each other at a 45° angle.

Periscopes are also used in submarines, so the submariners can see what is happening on the surface while the submarine is beneath the sea.

Periscopes

Watching a golf match is not always easy. It is often impossible to see the players because people stand in front of you. A **periscope** could help you see over the heads of other people.

The diagram below shows how periscopes can be made using just two plane mirrors. Light from distant objects strikes the top mirror at an angle of incidence of 45°. The angle of reflection is therefore 45°, so the light ray is turned through 90° and travels down the periscope tube. At the bottom of the periscope tube, the light strikes another mirror and is again turned through 90° where it can be seen by the observer.

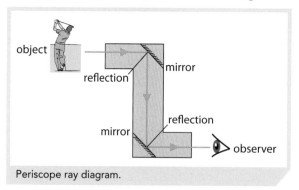

Periscope ray diagram.

Watching golf using periscopes.

Activity A

1. List the equipment you would need to build a periscope.
2. Describe how you would build your periscope.
3. Ask your teacher to check your plan, then build a periscope.
4. Evaluate your design: how well did the periscope work? Could you make a better one if you had more time?

Reflecting telescopes

Mirrors can also be used in telescopes to view distant stars and galaxies. Telescopes using mirrors are called **reflecting telescopes**.

The diagram below shows a ray diagram for a reflecting telescope. Light from the stars enters the telescope and is reflected by a concave mirror. The light rays converge towards a plane mirror, which reflects the light towards the eyepiece lens. The lens helps to magnify the image.

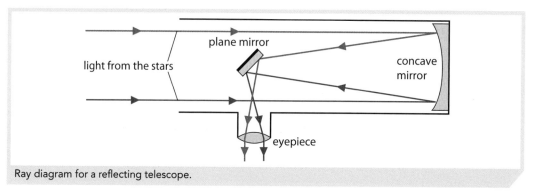

Ray diagram for a reflecting telescope.

Some telescopes only use lenses. These are called **refracting telescopes**. When astronomers study the stars, their telescopes need to gather as much light as possible so they can see dim and distant objects. It is easier to make very big mirrors than to make very big lenses, so most telescopes used for astronomy are reflecting telescopes.

Link

You will learn about lenses in lesson 6.15.

Case study

The James Webb Space Telescope (JWST), which will replace the Hubble Space Telescope, is being constructed by the USA and Europe, including the UK. The Royal Observatory in Edinburgh is involved in this project. The JWST will look at the earliest stars formed after the Big Bang as well as looking for other planetary systems that might support life. It is due to be launched in 2018.

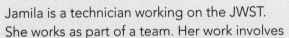

Jamila is a technician working on the JWST. She works as part of a team. Her work involves developing materials that can withstand the very cold temperatures in space. Materials behave differently at low temperatures. A material that is elastic at room temperature can be brittle at low temperatures. Jamila is also involved in the construction of some of the mirrors. These mirrors are coated with thin films of gold, which is smoother than ordinary glass.

1 Why do you think gold films are better than ordinary glass for mirrors?

2 Electric wires in homes are insulated using a type of plastic. Why do you think this plastic may not be suitable for use on the JWST?

3 Which type of telescope do you think the JWST will be?

Did you know?

The Hubble Space Telescope is a reflecting telescope which uses two mirrors. Putting a telescope into orbit allows astronomers to make observations in space, where the light has not been distorted or absorbed by the Earth's atmosphere.

Hubble Space Telescope.

Assessment activity 6.8   2C.P7 (part) | 2C.M5 (part) | 2C.D4 (part)

You are an optical engineer, working to develop better mirrors to help HGV drivers see cyclists in their blind spots. Your manager has asked you to produce some teaching materials for new trainees in the company.

1 Use ray diagrams to explain how an image is formed in a plane mirror.

2 Draw ray diagrams to illustrate the following applications of plane mirrors:
  (a) rear view mirror in a car
  (b) periscope used in submarines.

3 Draw up a list of bullet points to describe the similarities and differences between the images formed by plane mirrors and by curved mirrors.

4 Explain how a reflecting telescope works. Include a ray diagram in your answer.

Tips

When drawing ray diagrams, make sure that the angle of incidence and angle of reflection are measured from the normal. Always use a ruler to draw straight lines, and draw angles accurately. Always label your diagrams when describing applications.

Just checking

1 Describe two situations in which a periscope might be used.

2 What type of mirror is used in a periscope?

3 What types of mirror are used in a typical reflecting telescope?

Lesson outcome

You should be able to draw ray diagrams for simple applications of mirrors.

6.15 Refraction

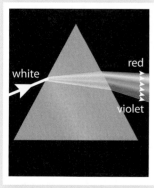

Refraction of light by a prism.

You make use of lenses every time you look at something or take a photo. Lenses work because light travels at different speeds in different materials.

Light slows down when it enters glass. This causes the path of the light to bend. This bending effect is called **refraction**. Refraction occurs whenever light enters or leaves transparent materials.

Refraction of light in lenses

Convex lenses are fat in the middle and thin at the edges. They are sometimes called converging lenses, because they refract parallel rays of light so they converge to a point. This point is called the **focal point**, F, and the distance between the focal point and the lens is called the **focal length**, f.

Concave lenses are thick at the edges and thin in the middle. They are sometimes called diverging lenses because they make rays diverge (spread out). The point from which the refracted rays *appear* to come is the focal point of a concave lens.

The sizes of objects and their images viewed through a lens are measured from an imaginary line through the centre of the lens called the **principal axis**.

Refraction of light in prisms

White light is a mixture of different colours of light. Each wavelength of light is refracted by a different amount when it passes into or out of a **prism**. Red is refracted the least and violet is refracted the most. This is why white light is split into the colours of the visible spectrum. This effect is called **dispersion**.

Refraction is also the reason rainbows are seen in the sky. In this case, light passes through water droplets, which refract the light and disperse it into its different colours.

Ray diagrams

Convex lenses are used in telescopes, in photographic cameras and in magnifying glasses. The type of image formed by a convex lens depends on the distance of the object from the lens.

Ray diagrams are drawn using three rays. The 'top' of the image forms where the three rays meet. The 'bottom' of the image is on the principal axis. When the object is more than one focal length away from the lens the image is real, **inverted** (upside down) and **diminished** (smaller than the object). A **real image** is one that can be projected on to a screen.

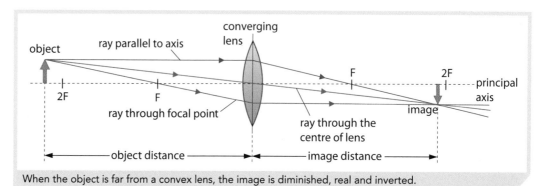

When the object is far from a convex lens, the image is diminished, real and inverted.

The image gets bigger the closer you get to F. At F, no image is formed. When the object is closer to the lens than the focal point, the image is a magnified, **virtual image**. This is how a magnifying glass works. A virtual image cannot be projected onto a screen.

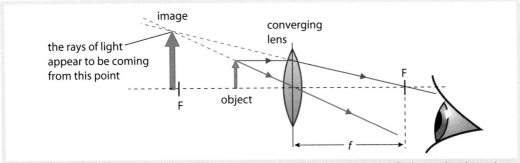

When the object is closer to a convex lens than the focal point, the image is magnified, virtual and upright.

Concave lenses are used in combination with convex lenses in some cameras and telescopes.

The ray diagrams below show how a concave lens works. The image formed by a concave lens is *always* virtual and **upright**, and is *smaller* than the object. The image gets smaller as the object is placed further away from the lens.

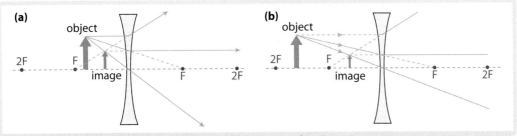

Concave lens: **(a)** object closer than F; **(b)** object further away than F.

Assessment activity 6.9

2C.P7 (part) | 2C.M5 (part)

You have applied for a job as a technical assistant at an optician's practice. Part of the interview task is to draw ray diagrams and interpret them correctly. You have been supplied with a 5-cm-tall light bulb that is placed 50 cm away from a convex lens that has a focal length of 15 cm. Using graph paper and a ruler, complete the following tasks.

1 Draw a ray diagram to scale, labelling the object, focal point and image.
   (a) How tall is the image?
   (b) Is the image upright or inverted?
   (c) What is the nature of the image – is it virtual or real?
   (d) What happens to the image if the bulb is moved to be 10 cm from the lens?

2 You are instructed to replace the convex lens with a concave lens. Repeat the questions in **1** for the concave lens.

Just checking

1 What is meant by the focal length of a lens?
2 What is meant by a virtual image?
3 Which kind of lens can be used as a magnifying glass?

Activity A

There is a relationship between the distance of the object from the lens (*u*), the distance of the image from the lens (*v*) and the focal length (*f*) for thin convex and concave lenses.

$$\frac{1}{f} = \frac{1}{v} + \frac{1}{u}$$

1 An object is 30 cm away from a convex lens. The lens has a focal length of 20 cm. Calculate how far the image will be from the lens.

2 Explain what kind of image you will see.

Tips

For 2C.P7 and 2C.M5, you need to make sure that you have the three rays clearly shown on your diagram. You must also correctly label the object, image and focal point on the graph paper.

Lesson outcome

You should know how to draw ray diagrams for convex and concave lenses.

6.16 The human eye

Our eyes use convex lenses to focus light – but human eyes don't always work properly. Luckily, we can use lenses in spectacles or contact lenses to help correct many eye problems.

Get started ➡️

In groups of two, list some similarities and differences between the lens in your eye and the lens in a camera.

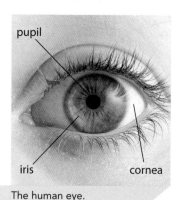
The human eye.

How the eye works

The retina is at the back of your eye and is the part of your eye that detects light. Light rays from the objects you are looking at are focused on the retina, so that your brain gets a clear image.

As light enters the eye it is refracted as it passes through the cornea and the lens so that an image is formed on the retina. Here, the image is converted to an electrical signal and transmitted via the optic nerve to the brain. Although the image is inverted, the brain has learnt to process the image so that it appears the right way up. Muscles around the lens can make the lens fatter or thinner to focus on close or distant objects.

Some people suffer from defects in their vision because the lens in the eye cannot change shape enough to focus an image on the retina. To overcome this, people wear spectacles or contact lenses, or have laser eye surgery.

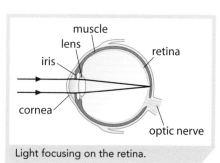
Light focusing on the retina.

Short and long sight

Short sight is a common defect in which a person can focus on close objects but cannot see distant objects clearly. The lens refracts light too much when looking at distant objects, and light is brought to a focus in front of the retina.

A concave lens can correct short sightedness.

If someone is long sighted they can see distant objects clearly but close objects will look blurred. This is because their lens does not refract light from close objects enough to focus the image on the retina.

Long sight can be corrected by using a convex lens – this is how reading glasses work.

Long sight

Convex lenses correct long sight.

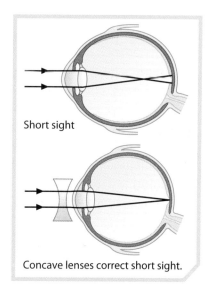
Short sight

Concave lenses correct short sight.

Link 🔄

Lesson 6.15 explains how concave and convex lenses work.

Worked example ✏️

You are a trainee optician working in the optical unit of a local hospital. Part of your work is to provide an eye test for people with eye defects. To do this you have been asked to model the eye as a convex lens, and to use these two equations to answer the following questions.

$$\frac{1}{f} = \frac{1}{v} + \frac{1}{u}$$

$$\frac{h_i}{h_o} = \frac{v}{u}$$

u = distance of object from lens
v = distance of image from lens
f = focal length
h_o = height of object
h_i = height of image

Your patient is looking at a test chart that is 1.75 m high and 2.74 m away. The focal length of the patient's eye is 0.0179 m.

1 Calculate the distance of the image from the eye lens.

Step 1 You know u and f; you need to find v.

Use the equation: $\dfrac{1}{f} = \dfrac{1}{v} + \dfrac{1}{u}$

Step 2 Rearrange to make v the subject: $\dfrac{1}{v} = \dfrac{1}{f} - \dfrac{1}{u}$

Step 3 Substitute values: $\dfrac{1}{v} = \dfrac{1}{0.0179\,\text{m}} - \dfrac{1}{2.74\,\text{m}} = 55.5$

$$v = \frac{1}{55.5} = 0.018\,\text{m}$$

2 Calculate the height of the image.

Step 1 Use the equation: $\dfrac{h_i}{h_o} = \dfrac{v}{u}$

Step 2 Rearrange to make h_i the subject: $h_i = \dfrac{v}{u} \times h_o$

Step 3 Substitute values: $h_i = \dfrac{0.018\,\text{m}}{2.74\,\text{m}} \times 1.75\,\text{m} = 0.0115\,\text{m}$

3 What is the nature of the image that is formed – is it virtual or real? Explain your answer.

The image formed is real because the object is further away than the focal length.

A chart like this one is used in an eye test.

Activity A

You have extended the experiment in the Worked example by asking the patient to stand at different distances, and have calculated the height of the image expected. You have assumed that the focal length of the eye is fixed at 1.79 cm. Your data are shown in the table.

Object distance (u) (cm)	Image distance (v) (cm)	Height of image (h_i) (cm)
150	1.81	2.11
350	1.80	0.90
950	1.79	0.33

1 What relationship do the data show between where the patient stands and the calculated height of the image? Comment on the quality of the image that would be formed.

2 The data show that the image distance changes with object distance. Explain why this is a problem for the eye and how the eye overcomes this. You may need to do further research for this!

Just checking

1 What type of lens is in our eyes?

2 What can a long-sighted person see clearly?

3 How can someone who is short sighted have their vision corrected?

Lesson outcomes

You should know how the lenses in your eyes work and how optical lenses can be used to correct simple eye problems.

6.17 Total internal reflection

Key terms

Optical fibre – Flexible transparent fibre that transmits light using total internal reflection.

Total internal reflection – Occurs when light passes from one material to another and the angle of incidence reaches a critical size, so that refraction is not possible and all the light is reflected.

Did you know?

The critical angle for glass is 42° while for water it is 49°.

A surgeon is able to carry out sophisticated laser surgery on a patient because of our understanding of how light behaves when it passes from glass to air.

When light travels from a material such as glass or water into air, it is refracted away from the normal as it leaves the material. However, if the angle of incidence at which the light strikes the glass/air interface is greater than a **critical angle**, then the light is reflected back inside the glass. The edge acts like a mirror. This is known as **total internal reflection**. The critical angle depends on the material the light is travelling through.

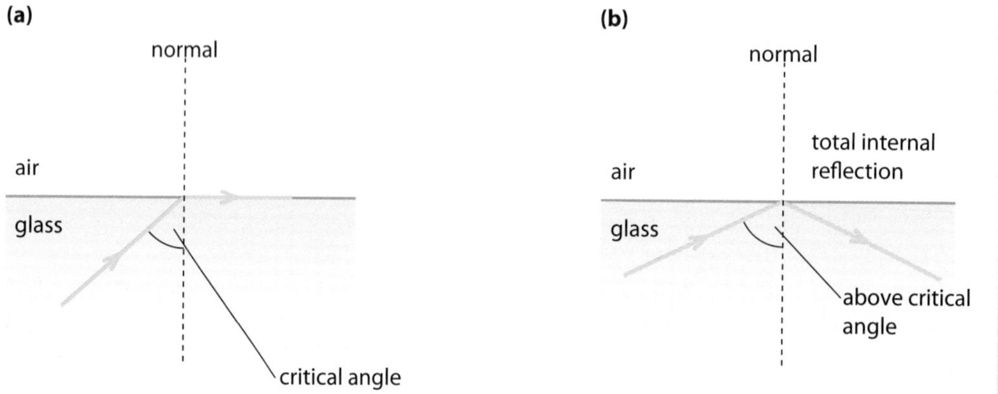

(a) Light leaving glass at the critical angle. **(b)** Total internal reflection.

Total internal reflection in action

It is important that car drivers can see cyclists at night. So, for safety, bicycles should be fitted with reflectors that reflect the light from car headlights.

A plane mirror would not reflect light from headlights back towards the driver's eyes. Bicycle reflectors contain many tiny plastic prisms. Light coming from a car is internally reflected twice by the prisms so that it is sent back towards the car. This alerts the driver of the car to the cyclist's presence.

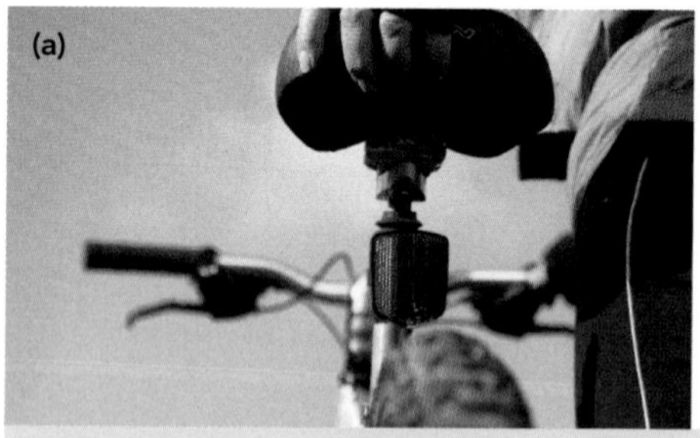

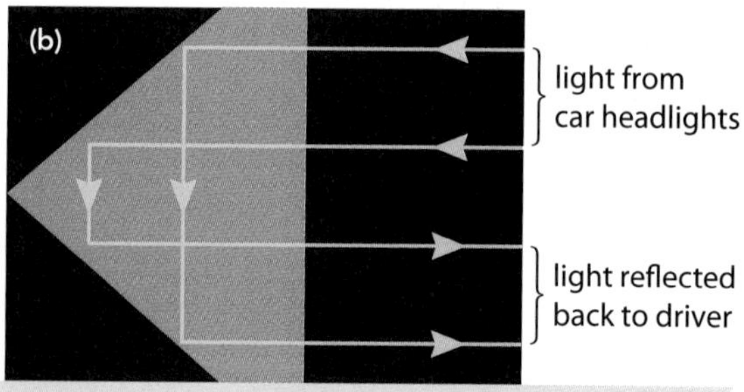

(a) Bicycle reflector and **(b)** ray diagram for bicycle reflector.

Activity A

A simple telescope produces an inverted image. Binoculars with the same magnification as a telescope are much shorter and produce an upright image. Find out how prisms are used in binoculars, and draw a ray diagram to illustrate your findings.

Optical fibres

Optical fibres use total internal reflection to guide light through a material. An optical fibre is normally very thin, with a core (centre) made of very clear glass. The core is surrounded by a different glassy material (cladding) that causes rays of light within the core to be totally internally reflected. The edges of the core act like perfect mirrors and stop light escaping from the fibre.

Optical fibres are used in telecommunications and computer networking. They are also used in many areas of medicine, including keyhole surgery. Doctors use optical fibres to look inside the human body. If, for instance, an ulcer is detected in a patient, a surgeon can send a laser beam down the optical fibre to burn the ulcer.

Light path in an optical fibre.

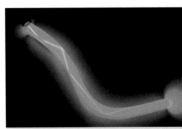

Total internal reflection in optical fibres. Notice the green laser light reflecting from one surface to another.

Activity B

Optical fibres are surrounded by a material called cladding, which stops light escaping from the fibre. If some light did escape, it would mean information would be lost and image quality would be distorted.

1 Find out what types of cladding are used in optical fibres.

2 Using the words critical angle and total internal reflection, explain how cladding works.

Assessment activity 6.10 | 2C.P7 (part) | 2C.D4 (part)

You are a junior technician in a materials science laboratory. You are working on devising ways of using optical materials in various applications. There is going to be an open day at your laboratory, and you have been asked to produce a poster to explain your work.

1 Draw ray diagrams to explain how prisms can be used in
  (a) binoculars and
  (b) bicycle safety reflectors.

2 Use your knowledge of how light travels in glass to explain how optical fibres work and how they are used in keyhole surgery.

Tips

For 2C.P7, remember that the angle of incidence/reflection is measured from the normal, and your diagram should clearly show this.

For 2C.D4, the key is to explain how the light travels and not concentrate on what the application does. Some description of critical angle and total internal reflection is expected, with ray diagrams.

Just checking

1 What is a critical angle?

2 How can you use a prism to split up white light?

3 State two applications of total internal reflection.

Lesson outcome

You should be able to explain how total internal reflection of light can be used in applications.

6.18 Investigating sound

Get started

How many people do you know who have noisy neighbours? Discuss where in the house you think you are more likely to hear your neighbours. Is there anywhere in the house where you can't hear your neighbours, and do you know why that is?

Key terms

Compression – Area where molecules are closer together. This will be an area of high pressure.

Medium – A material, such as air, water or a solid, that sound travels through.

Rarefaction – Area where molecules are spread out. This will be an area of low pressure.

You are watching a science fiction film – there is a battle and a spaceship blows up with a flash of light and a sound like an explosion. But if you were really watching a space battle, there would be no sound at all! Sound cannot travel through empty space.

What is sound?

Sound is produced whenever an object vibrates. The vibrations travel through a surrounding **medium** such as air, water or a solid material. At home you hear the TV because the speakers in the TV vibrate and these vibrations are passed on by air particles.

The direction of a sound wave is in the same direction as the vibrations. It is a longitudinal wave. The diagram below shows how a sound wave travels in air.

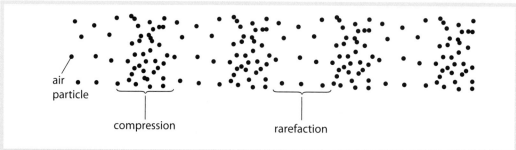

air particle

compression   rarefaction

The different spacing between the air particles shows how a sound wave travels in air.

Sound in air is propagated (transmitted, or sent on) by air particles pushing each other to and fro. This produces regions where the molecules are close together – called **compressions**. The air pressure will be higher where the particles are closer together. There are also areas where the molecules are further apart – called **rarefactions** – where the air pressure will be lower than normal. Sound waves spread out in all directions from the vibrating object.

Sound and materials

Sound travels at different speeds in different materials, because of the different ways in which the particles are arranged. The table shows the speed of sound in air, water, brick and aluminium.

Material	Speed (m/s)
Air	330
Water	1433
Brick	4178
Aluminium	6420

Sound travels almost 20 times faster in aluminium than air. This is because the atoms are more closely packed in a metal and so more atoms are able to vibrate and transmit sound energy.

Soft, squashy materials do not pass on vibrations well. They absorb the energy being transferred by the sound waves. This is why the noise you make in a classroom, with hard floors and usually no curtains, sounds louder than if you made the same noise in a room with carpets and curtains. Other useful materials for providing sound insulation are foam rubber and felt.

As the aircraft reaches the speed of sound, the speed at which sound waves travel, a shockwave forms around the aircraft. Water vapour in the low pressure area behind the shockwave condenses, forming the cloud which you can see here.

Activity A

You and another person are simulating sound using a Slinky® spring. You see the patterns shown below.

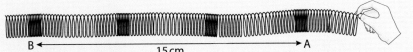

B ←————— 15 cm —————→ A

1 Copy the diagram of the Slinky®. Identify the compressions and rarefactions in the Slinky® spring by labelling these regions on the diagram.

Use your answers to question **1** to help you to answer questions **2**, **3** and **4**.

You will also need to use knowledge from Principles of Applied Science lesson 1.27, and the equation:

$$\text{speed} = \text{frequency} \times \text{wavelength}$$

2 What is the wavelength of the wave?

3 You and your partner measured the time taken for pattern A to reach B as 3 seconds. Calculate the frequency of the wave.

4 From your answers to questions **2** and **3**, calculate the speed of the wave.

5 Posters for the movie 'Alien' said 'In space no one can hear you scream'. Explain why this statement is true.

? Did you know?

The Slinky® toy is a spring that stretches and can bounce. It was invented by a naval engineer in the 1940s. He was working on springs that could give support to ship's instruments and he discovered the properties of the spring by accident.

Assessment activity 6.11 | 2C.P8 | 2C.M6

You are the sound engineer for a music company. With the aid of a labelled diagram, describe to your new trainee how sound is produced and how it travels.

Tips

For 2C.P8, remember to mention that sound needs a medium to travel through.

For 2C.M6, make sure that the terms molecule, rarefaction and compression are used on your labelled diagram, and that your description includes them too.

Just checking

You can make a model telephone using two plastic cups and some string. One person speaks into their plastic cup, and the other person holds their cup to their ear. The 'telephone' only works when the string is tight.

1 Explain how vibrations are passed from one person to the other by the 'telephone'.

2 Explain why the 'telephone' doesn't work if the string is not tight.

Lesson outcomes

You should be able to explain the need for a medium for the transmission of sound waves, and how compressions and rarefactions are formed in the medium.

6.19 Applications of sound

Get started

Sound is used in many applications. Discuss in groups three applications that use sound.

Key term

Ultrasound – Sound that is too high for humans to hear.

Did you know?

Bats use ultrasound waves to hunt and to prevent them flying into obstacles at night. The ultrasound waves bounce off anything in the air, and the echoes which are produced allow the bats to work out where the prey or obstacle is.

Did you know?

Ultrasound is also used to clean teeth and medical instruments.

Voice recognition

If someone you know speaks to you, you can usually tell who it is without seeing them. When you speak, the sound you make is a mixture of lots of different wavelengths of sound. Different people produce slightly different combinations of wavelengths.

Voice recognition can be used for security, so that only certain people are allowed into a building or room. It can also be used by police forces in solving crimes.

A computer converts sound waves into a digital 'fingerprint' (a waveform) and matches it to a stored sample. A typical audio waveform before the voice is digitised is shown above.

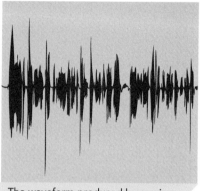

The waveform produced by a voice.

Ultrasound imaging

Sound waves above 20 kHz can't be heard by humans and are called **ultrasound**. Ultrasound is used in hospitals to form images of babies before they are born, by 'seeing' them in the mother's womb. The diagram below shows how this works. Using ultrasound is much safer for the baby and mother than using X-rays. Ultrasound images are also used to diagnose other medical conditions.

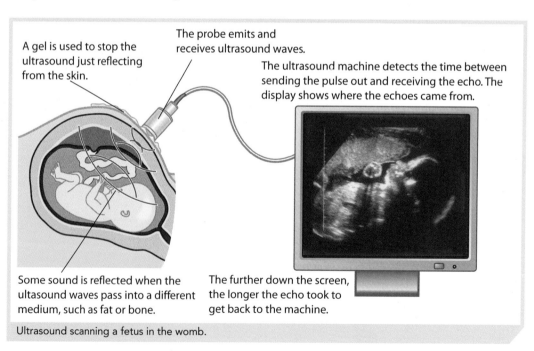

A gel is used to stop the ultrasound just reflecting from the skin.

The probe emits and receives ultrasound waves.

The ultrasound machine detects the time between sending the pulse out and receiving the echo. The display shows where the echoes came from.

Some sound is reflected when the ultrasound waves pass into a different medium, such as fat or bone.

The further down the screen, the longer the echo took to get back to the machine.

Ultrasound scanning a fetus in the womb.

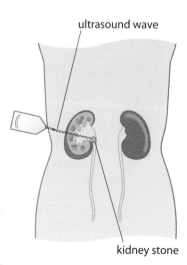

ultrasound wave

kidney stone

Ultrasound can be used to break up kidney stones.

Kidney stones are small lumps that form in the kidneys; they can cause pain and discomfort to patients. Rather than being removed by surgery, a powerful beam of ultrasound can be used to break them up into tiny pieces. These pieces are then expelled from the body with the urine.

1 What is the difference between ultrasound used to form images of babies in the womb and ultrasound used to break up kidney stones?

2 How do you think the ultrasound waves break up the kidney stones?

Sonar

Sound waves can be reflected, just like light waves. A reflected sound is called an **echo**. Echoes can be useful, for instance when used in sonar on board ships and submarines. SONAR stands for SOund NAvigation and Ranging. It can be used to find out how deep the sea is, or to find schools of fish.

The diagram on the right shows how sonar is used to check the depth of water. The sonar transmitter sends a pulse of ultrasound downwards. This travels through the water, and some of it is reflected when it hits the sea bed. The sonar detects the returning echo. It uses the time between the pulse and the echo, and the speed of sound in water, to calculate the depth of the water.

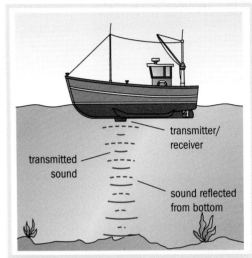

Sonar can be used to calculate the depth of water.

Assessment activity 6.12 | 2C.P8 | 2C.D5

1 In the form of a leaflet:
(a) describe how bats use sound in navigation
(b) describe how sound reflections are used in sonar
(c) describe how sound insulation is used to prevent noise in houses.

2 You have just been employed as a medical physics technician. In the form of a poster, explain how sound can be used in the following applications:
(a) ultrasound imaging
(b) ultrasound for use in treatment
(c) voice recognition as used in a security system to protect access to patient records.

Tips

The task for question **1** can be brief so don't get carried away. The main point here is to cover the three applications.

For 2C.P8, you need to link materials to how well sound travels through them.

For 2C.D5, make sure that you use scientific words in your description. Always refer to any diagrams you have drawn in your description.

Just checking

1 What is the difference between sound and ultrasound?

2 State two devices that use echoes.

Lesson outcome

You should know about everyday applications that use sound waves.

6.20 Building circuits

Get started

In groups of three or four, name some equipment an electrician might bring with them if they came to fix some electrical items at your home or school. What are the different pieces of equipment used for?

Key terms

Parallel circuit – Circuit in which electrical components are connected so that the same voltage is applied to each component.

Resistor – An electrical component that limits current flow.

Series circuit – Circuit in which electrical components are connected so that the same current flows through each component.

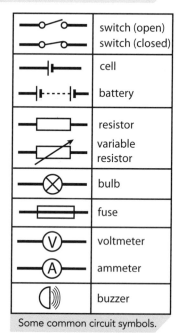

⊸o o⊸	switch (open)
⊸o͞o⊸	switch (closed)
⊣⊢	cell
⊣⊦----⊦⊢	battery
▭	resistor
⟋▱	variable resistor
⊗	bulb
▭	fuse
Ⓥ	voltmeter
Ⓐ	ammeter
⍉⟫	buzzer

Some common circuit symbols.

Link

Look at Principles of Applied Science lesson 3.7 to read more about electricity and circuits.

If you look inside a television or other electrical machine, the circuits look very complicated. However, all circuits are combinations of much simpler **series** and **parallel** circuits. In a series circuit (diagram **a**) all the components are on the same 'loop' of wire. In a parallel circuit (diagram **b**) there are several different paths that the current can follow.

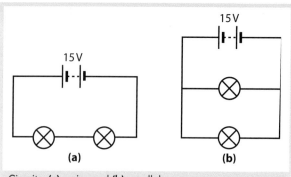

Circuits: **(a)** series and **(b)** parallel.

Measuring current and voltage

When a battery or power supply is connected to a circuit, charge will flow along the wire. The amount of charge passing each second is the current. Current is measured using an **ammeter**. Ammeters are always connected in series.

The amount of energy that each bulb in a circuit uses depends on the current and the voltage. The voltage is the amount of energy transferred by each unit of charge. Voltage is measured using a **voltmeter**. Voltmeters are always connected in parallel.

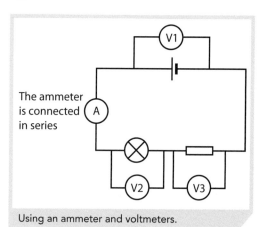

Using an ammeter and voltmeters.

The circuit diagram on the right shows three voltmeters. V1 is measuring the voltage provided by the cell (battery), V2 is measuring the voltage across the bulb, and V3 is measuring the voltage across the **resistor**.

In modern scientific laboratories and workshops, a multimeter is used to measure both voltage and current. Multimeters can also be used to measure resistance directly.

Investigating circuits

The diagrams below show a bulb and a resistor in **(a)** a series circuit and **(b)** a parallel circuit. As you can see, the currents and voltages are different in the two circuits.

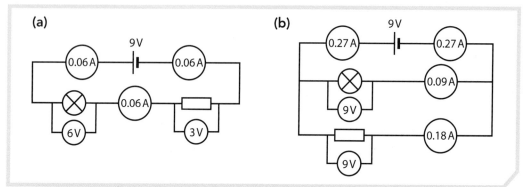

Rules for series circuits

Current: The current is the same everywhere in a series circuit.

Voltage: The voltage of the supply is divided between the different components, depending on the resistance of each component.

Rules for parallel circuits

Current: The current from the cell splits up when it reaches a junction. The proportion of the current that flows through each branch of the circuit depends on the resistances of the components in that branch.

Voltage: The voltage is the same across each branch of the circuit.

A digital multimeter being used to check the resistance of a motorcycle alternator.

Worked example

Use the circuit rules to help you to work out the answers to the questions below.

1 Resistor R and bulb B2 are like a mini-series circuit. What is the total voltage across R and B2?

6 V, because the voltage across each branch of a parallel circuit is the same as the voltage of the supply.

2 The current flowing through the cell is 0.6 A. The current flowing through R is 0.15 A.
   (a) What is the current flowing through B2?

   0.15 A, because the same current must flow through the resistor and B2.
   (b) What is the current flowing through bulb B1?

   0.45 A, because the currents flowing through each branch of the circuit must add up to the current through the cell.

3 The voltage across B2 is 2 V.
   (a) What is the voltage across R?

   4 V, as the total voltage across this branch is 6 V.
   (b) What does this tell you about the resistance of R compared to the resistance of B2?

   R has a higher resistance than B2, as it has a higher proportion of the voltage.

4 What can you say about the total resistance of R and B2 compared to the resistance of B1?

There is a higher current flowing through B1, so its resistance must be lower than the total resistance of R and B2.

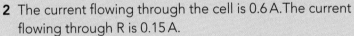

6 V
B1
B2
R

Copy these circuit diagrams and fill in the missing values. The two bulbs are identical.

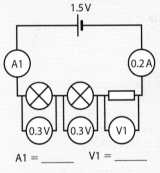

1.5 V
A1   0.2 A
0.3 V   0.3 V   V1

A1 = _____   V1 = _____

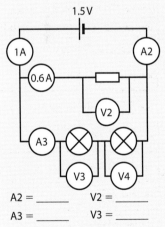

1.5 V
1A   A2
0.6 A
V2
A3
V3   V4

A2 = _____   V2 = _____
A3 = _____   V3 = _____
                V4 = _____

Just checking

1 What does an ammeter measure?
2 How would you connect an ammeter in an electrical circuit?
3 How would you connect a voltmeter in an electrical circuit?
4 Describe some differences between series and parallel circuits.

Lesson outcomes

You should understand the difference between series and parallel circuits, and how to connect voltmeters and ammeters to measure voltage and current across circuits.

6.21 Ohm's law

Get started

Discuss in groups what resistance means, and how the resistance of components affects the size of a current in a circuit. What else might affect the size of the current?

Key term

Ohm's law – States that, at constant temperature, the current through a conductor is directly proportional to the voltage across it.

Remember

You can use a 'magic triangle' to help you rearrange equations like that used in Ohm's law.

Step 1 If you want to find the resistance of a conductor and you know the voltage across it and the current through it then put your finger over the resistance (*R*).

Step 2 You can see $\frac{V}{I}$ in the triangle, so the equation you need is $R = \frac{V}{I}$

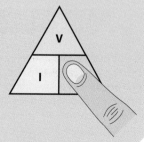

If you open any consumer electronics product, such as a radio, TV or mobile phone, you will find that there are many resistors inside. These resistors are important to control the current through the different circuits.

For most resistors and wires, if you double the voltage across it, the current flowing through it doubles. **Ohm's law** states that, provided the temperature remains constant, the current through a **conductor** is directly proportional to the voltage across it. That is, resistance is constant at constant temperature.

The relationship between voltage, current and resistance is given by this equation:

voltage (V) = current (A) × resistance (Ω)

$$V = I \times R$$

Ohm's law is represented in a graph by a straight line, as shown below.

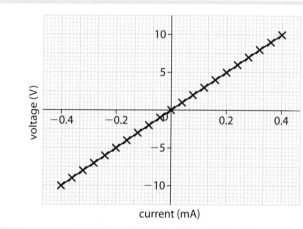

A graph of voltage versus current produces a straight line, as predicted by Ohm's law. The resistance can be obtained by finding the gradient of the graph (*R* = gradient).

Worked example

A resistor is part of an electrical circuit. The current through the resistor was measured to be 0.01 A and the voltage was measured to be 3.2 V. Use the equation that describes Ohm's law to calculate its resistance.

Step 1 Write down the equation: voltage = current × resistance

Step 2 Rearrange the equation to make resistance the subject (use the triangle to help you):

$$\text{resistance} = \frac{\text{voltage}}{\text{current}}$$

Step 3 Substitute for voltage and current: resistance = 3.2 V ÷ 0.01 A = 320 Ω

Activity A

Use the graph of voltage vs current to help you answer the following questions.

1 What voltage produces a current of 0.20 mA?

2 Use the graph to work out the resistance.

3 Copy the graph and draw another line that shows the results for a resistor with a smaller resistance. Label that line X.

Activity B

1 Two resistors (R1 and R2) are connected in series with a 230 V mains supply. The voltage across R1 is measured to be 80 V. Calculate the voltage across R2.
2 If R1 has a resistance of 100 Ω, calculate the current through R1 that would be recorded by an ammeter.
3 Use your answer from question **2** to work out the resistance of R2.

Assessment activity 6.13 | 2D.P9 | 2D.M7 |

Mimi is an assistant technician and she has asked you to help her carry out some electrical measurements.

1 She asks you to draw a circuit having a battery and two unknown resistors, as well as the meters that are going to be used to measure current and voltage. The resistors are to be connected in series.
2 Now you need to draw another circuit with two different resistors. The resistors are to be connected in parallel. The current and the voltage across each resistor will also need to be measured.
3 Mimi tabulates her results as shown below. Calculate the resistance of each resistor for both the series and the parallel circuit. Remember: 1000 mA = 1 A, so to convert from mA to A you must divide by 1000.

Circuit	Current through R1 (mA)	Current through R2 (mA)	Voltage across R1 (V)	Voltage across R2 (V)
Series	2.2	2.2	5.2	3.8
Parallel	1.2	5.8	9.0	9.0

4 The total resistance of a series circuit is equal to the sum of the values of the individual resistors: $R_{total} = R1 + R2$
Work out the total resistance for Mimi's series circuit.

If you have time, you could carry out these experiments yourself and use your own results in the calculations.

Tips

Make sure your diagrams show a complete circuit.

For 2D.P9, make sure that the voltmeter and ammeter are shown to be connected correctly and that you have used the correct symbols.

For 2D.M7, the final answer needs to have the correct unit for resistance, otherwise you will not achieve this criterion.

Just checking

1 What is the condition needed for Ohm's law to be obeyed by a resistor?
2 What quantities do you need to measure to allow you to calculate the resistance of a conductor?

Did you know?

Ohm's law is named after the German physicist Georg Ohm who found, in 1827, that there was a direct relationship between the voltage across a conductor and the current flowing through it, at constant temperature.

Georg Ohm, 1789–1854.

Lesson outcome

You should be able to use Ohm's law to calculate resistance.

6.22 Components that don't obey Ohm's law

Key term

Light-dependent resistor (LDR) – A component with high resistance in the dark and low resistance in the light.

Did you know?

Filament bulbs are very inefficient. Less than 10% of the energy they use is turned into visible light – the rest is turned into heat. This is why filament light bulbs are being phased out, to be replaced by low energy fluorescent bulbs.

Although resistors are important components in many electrical devices, there are many everyday appliances that use components that don't obey Ohm's law.

Filament bulbs are still used in lighting applications such as in cars and decorations. Currents pass through the filament differently from resistors. Understanding this behaviour is important to keep circuits safe.

Filament bulbs

The photo below shows a light bulb with a filament. The filament gets hot enough to glow when a current flows through it. The diagram shows a circuit that can be used to investigate how the current through the bulb changes when the voltage changes. The graph shows the results of such an investigation. You will see that the line on the graph is not straight – showing that the lamp doesn't obey Ohm's law. This is because the current heats the filament, causing the resistance of the filament to increase. As the voltage is increased, the current also increases and makes the filament hotter. The resistance continues to increase as the temperature rises. Each additional volt produces a smaller and smaller increase in current.

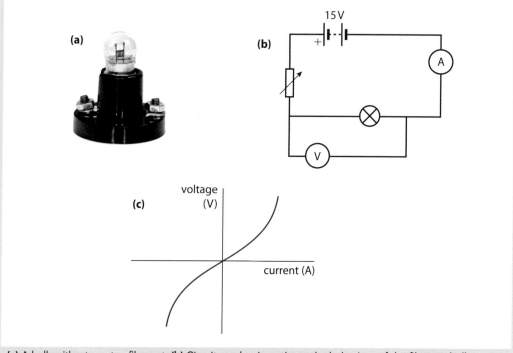

(a) A bulb with a tungsten filament. (b) Circuit used to investigate the behaviour of the filament bulb.
(c) Graph of voltage against current for the filament bulb.

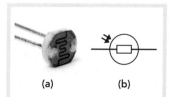

(a) A picture of an LDR (light-dependent resistor), and (b) LDR circuit symbol.

Light-dependent resistors (LDRs)

A **light-dependent resistor** (LDR) is a special resistor which changes its resistance as light shines on it. LDRs can be used, for example, in security lights that come on when it gets dark.

The diagram below shows the equipment needed to investigate what happens to the resistance of an LDR as light of varying brightness shines on it.

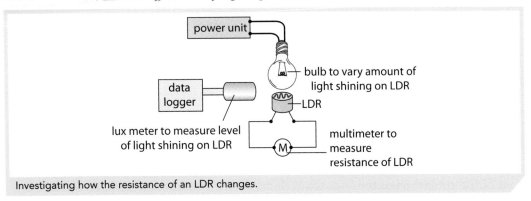

Investigating how the resistance of an LDR changes.

When the brightness of the light is changed, the lux and resistance also change, as shown in the graph below.

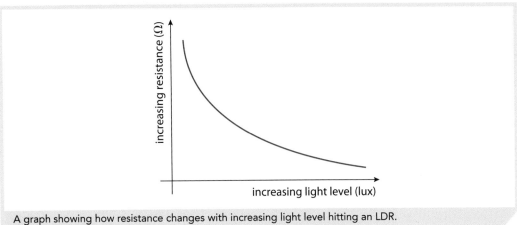

A graph showing how resistance changes with increasing light level hitting an LDR.

With no light on the LDR (in the dark), the resistance of the LDR is over 65 kΩ. Some LDRs can have a resistance as high as 1 000 000 ohms (1 MΩ) in the dark.

The circuit diagram on the right shows the circuit used to investigate how the current changes when the voltage changes for an LDR. You would also need a light source connected to a power supply unit, to vary the amount of light shining on the LDR, and a lux meter to measure the level of light shining on the LDR.

The voltage is changed to different values using a variable resistor. For each voltage, the current is recorded with the light source off, and when the light source is switched on. The temperature is kept constant.

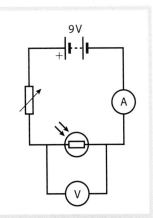

The circuit used to measure the voltage–current characteristics of an LDR.

This security light comes on when it gets dark.

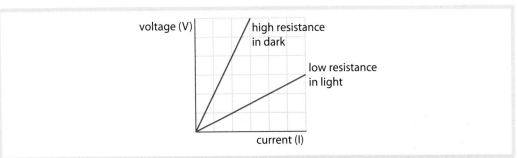

A voltage–current graph for the LDR in the dark and in the light. Notice that the slope is higher in the dark than in the light, showing that the resistance is higher in the dark.

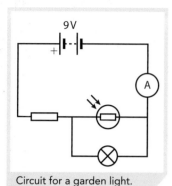

9V

Circuit for a garden light.

A solar-powered garden light. In solar-powered lights, energy from the sun is stored in batteries until it is needed.

Using LDRs

An LDR can be used in a circuit that controls a small garden light, as shown on the left.

The first thing you will notice about this circuit is that a resistor is connected in series to an LDR. The LDR is connected in parallel to the bulb, which will be used as the garden light. To understand how this circuit works, you need to remember the rules for current and voltage in series and parallel circuits.

In the dark, the LDR has a high resistance, making it difficult for the current to flow through it. So most of the current flows through the bulb and the bulb lights up.

In the light, the LDR has a lower resistance, so the current flows through it more easily. Most of the current flows through the LDR instead of through the bulb, so the bulb doesn't light up.

Activity A

Imagine you have made a box for your pet tortoise to sleep in. Sometimes, the tortoise manages to open the hatch and escape from the box. This could be dangerous for your tortoise, because you have a pond that it may fall into.
You want to modify the garden light circuit, described above, to design an LDR circuit you can put inside the box so you will know when the tortoise has escaped. The bulb is replaced by a buzzer.

1 How would you modify the position of the LDR to make the buzzer work when light falls on the LDR?

2 Explain how your modified circuit will work.

Assessment activity 6.14 | 2D.D6

1 Filament bulbs used to be used for the majority of lighting fixtures at home. Plan an experiment to investigate the voltage–current characteristics of a filament bulb.

2 Describe how the current of the bulb will vary with voltage.

3 Analyse why the current behaves in this way by describing how the resistance changes.

Tips

Make sure you describe how the component works in the circuit, not just what the circuit achieves.

For 2D.D6, your description of the filament bulb needs to relate to the shape of the voltage–current graph.

Lesson outcomes

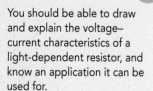

You should be able to draw and explain the voltage–current characteristics of a light-dependent resistor, and know an application it can be used for.

Just checking

1 Why does a filament bulb not obey Ohm's law?

2 What does the graph in the middle of page 97 tell you about how the resistance of an LDR changes with light level?

WorkSpace

▶ Jack

Electrician

I work at an electrical company. I normally go to work dressed in my uniform overalls, including shoe protectors. I always have my diagnostic pack and tools.

My day starts with a meeting of all electricians at our work site, so the head electrician can plan the day's activities with the rest of us. Sometimes we work on homes, other times it is large office blocks or factories.

Yesterday I was asked to extend the ring main of an industrial unit. I needed to link existing wiring to new wiring. This involves testing existing wires with continuity testers. I also tested the electrical circuit breakers to make sure they tripped within the safety time. I then placed labels onto the sockets. This is important so that future electricians will know that the sockets have been tested.

Yesterday I also had a job in a house. The owner wanted me to put a new feed into the kitchen and to the bathroom lights. Once I decided the type of cable needed I gave a quote to the owner for the job. Having agreed pricing, I went back to the work site to check our stock levels. If stock is low I have to go to the local warehouse to buy any items I need.

Every day my job involves working with different people and in different places. As well as my knowledge of electricity and safety, I also have to be aware of stock control, and be able to order equipment and stock at competitive prices.

Think about it

1 What kind of electrical instruments will Jack have in his diagnostic pack?

2 Why do you think Jack needs to wear shoe protectors in his job?

3 What do you think would happen if the trip takes longer than the safety time?

4 What other skills do you think Jack has to enable him to do his job well?

6.23 Thermistors

Key term

Negative temperature coefficient (NTC) thermistor – An electrical component that reduces its resistance as it gets warmer.

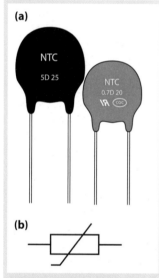

A thermistor **(a)** and its circuit symbol **(b)**.

When engineers design heating systems for houses and places of work, they don't just rely on resistors for controlling current flow. They use components such as **negative temperature coefficient (NTC) thermistors**. An NTC thermistor is another type of component that doesn't obey Ohm's law – its resistance decreases as it gets hot.

The diagram on the right shows the equipment needed to investigate how the resistance of a thermistor depends on temperature.

The multimeter is used to measure the resistance of the thermistor at several different temperatures.

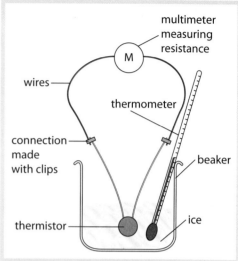

Apparatus to investigate how the resistance of a thermistor varies with temperature.

The voltage–current characteristics of a thermistor can be investigated using the circuit shown in diagram **(a)** below. The temperature will be kept constant. The voltage is changed by altering the setting on the power supply. Graph **(b)** shows that the thermistor does not obey Ohm's law.

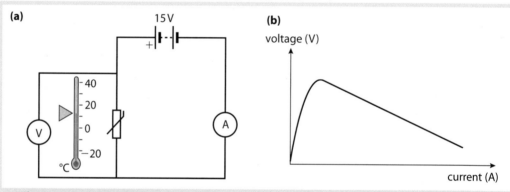

(a) The circuit used to investigate the voltage–current characteristics of a thermistor. **(b)** A voltage–current graph for a thermistor at a constant temperature. Notice that the graph does not obey Ohm's law.

Activity A

Use this graph showing resistance of a thermistor vs temperature to answer the following questions.

1 What is the resistance at 10 °C?

2 What is the temperature when the resistance is 10 000 Ω?

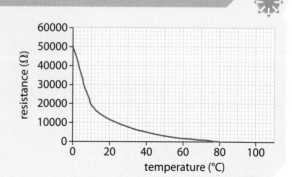

Using thermistors

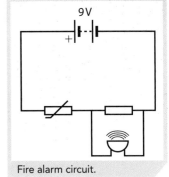

Fire alarm circuit.

The diagram on the right shows how a thermistor can be used as part of a fire alarm system.

When there is a fire and the temperature increases, the resistance of the thermistor decreases, so the thermistor gets a smaller proportion of the voltage from the supply – voltage rule for parallel circuits. This means the voltage across the resistor increases.

The buzzer is connected in parallel to the resistor, so the voltage across the buzzer also increases. If the temperature is high enough, it provides enough voltage to switch the buzzer on.

If the temperature falls again, the resistance of the thermistor increases and so does the voltage across it. The voltage across the resistor and the buzzer both decrease, and the buzzer switches off.

Assessment activity 6.15 | 2D.P10 | 2D.M8 | 2D.D7

You are an electrical engineering technician working with thermistors.

1 You have been instructed to investigate the use of a thermistor as part of a greenhouse alarm.

   (a) List what measurements you will need to take.

   (b) Plan how the thermistor would be used.

   (c) Check your plan with your teacher, then carry out your investigation.

2 You have just completed setting up a temperature control system for fridges for the National Blood Service. This is important to ensure that the blood is kept at specific temperatures. Your investigation produced the following data from a circuit that used a thermistor system as in the fire alarm circuit diagram.

Temperature (°C)	Current (mA)	Voltage (V)	Resistance (Ω)
−20	1.77	13.2	
−10	2.85	12.1	
0	4.10	10.9	
10	5.86	9.14	
20	7.53	7.47	
30	8.94	6.06	
40	10.4	4.58	

   (a) Calculate the resistance of the thermistor for each temperature and plot graphs of current vs temperature and resistance vs temperature.

   (b) Use your graphs to write conclusions on how the resistance varies with temperature and how the current varies with temperature.

   (c) Use the graph to work out at what temperature the current would reach 5 mA.

   (d) Suggest how you could improve this circuit and use it in your greenhouse alarm so that it can also detect low light conditions.

 Tips

For the thermistor, remember that you need to measure voltage, current and temperature.

For 2D.P10, make sure your plan for the experiment is realistic and sets out exactly what you want to do.

For 2D.M8, the graphs need to have the axes labelled, with temperature on the x-axis. Don't forget to change the current into amps before calculating the resistances.

For 2D.D7, you need to provide supporting evidence for your conclusions.

 Just checking

1 There are different types of thermistor. In this lesson you have learnt about the NTC type. What does NTC mean?

2 How does the resistance of an NTC thermistor change when its temperature increases?

 Lesson outcomes

You should be able to draw and explain the voltage–current characteristics of an NTC thermistor, and know an application it can be used for.

Introduction

In this unit, you will gain knowledge and skills that may enable you to embark on a career in biology, health care, nutrition, nursing, childcare, health promotion or medical laboratory services.

You will be able to develop your knowledge of health-related issues that you met in Unit 4. You will also build on your knowledge of the core science concepts you studied in Unit 1 in the Principles of Applied Science Award.

You will look at aspects of diet and exercise, and consider how the government tries to improve the nation's health.

You will examine some of the screening programmes offered by the National Health Service (NHS) to try to diagnose health problems before people get symptoms or become ill.

You will see how your body defends itself against infections, look at the advantages and disadvantages of vaccination and learn about some treatments using antimicrobial and analgesic drugs. You will also find out about blood and organ donation, and stem cell therapy.

Assessment: You will be assessed using a series of internally-assessed assignments.

Learning aims

After completing this unit you should:

a have investigated factors that contribute to healthy living

b know how preventative measures can be used to support healthy living

c have investigated how some treatments are used when illness occurs.

" I really enjoyed this unit. I hadn't realised that we can do so much to improve our health by eating properly and not smoking or drinking too much alcohol, and I didn't know that alcohol increases the risk of getting cancer. I found the section on stem cell therapy very interesting – one day we might be able to grow spare organs for when we need them.

Hannah, *15 years old*

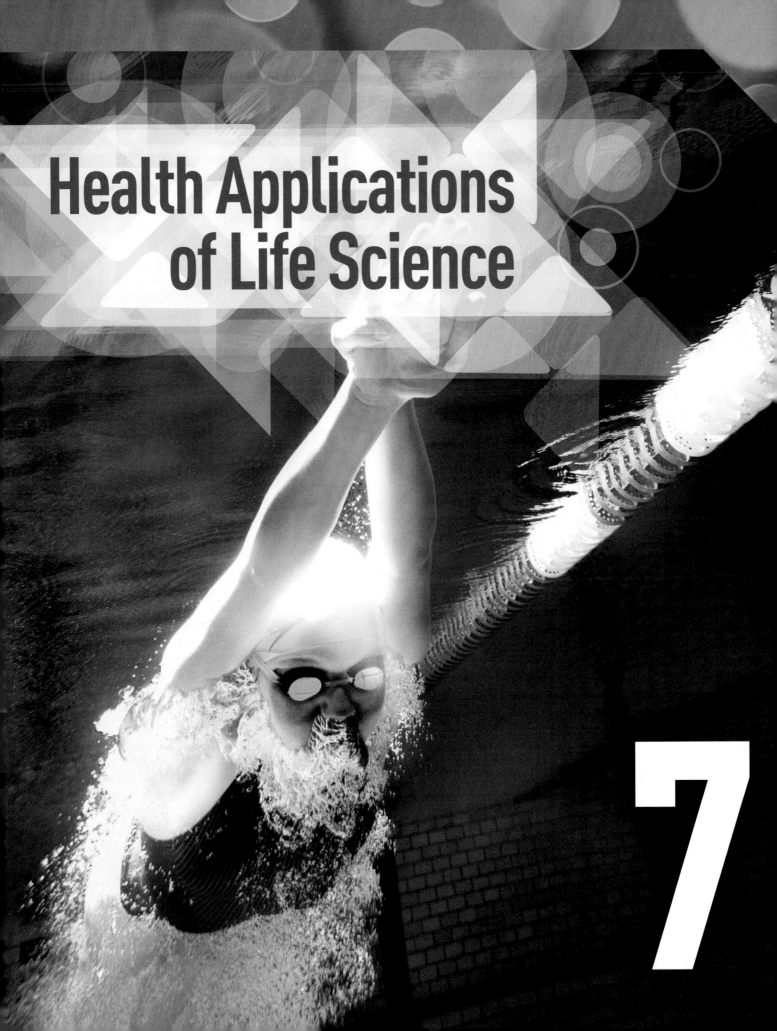

Health Applications of Life Science

7

BTEC
Assessment Zone

This table shows you what you must do to achieve a Level 1 Pass, or a Level 2 Pass, Merit or Distinction grade, and where you can find activities in this book to help you.

Assessment criteria			
To achieve a Level 1 Pass grade, the evidence must show that you are able to:	To achieve a Level 2 Pass grade, the evidence must show that you are able to:	To achieve a Level 2 Merit grade, the evidence must show that you are able to:	To achieve a Level 2 Distinction grade, the evidence must show that you are able to:
Learning aim A: Investigate factors that contribute to healthy living			
1A.1 Explain the importance of a balanced diet and exercise. Assessment activity 7.1	**2A.P1** Describe the possible effects of diet and exercise on the functioning of the human body. Assessment activity 7.1	**2A.M1** Explain how the diet and exercise plan will affect the functioning of the human body. Assessment activity 7.1	**2A.D1** Evaluate the diet and exercise plan, and justify the menus and activities chosen. Assessment activity 7.1
1A.2 Identify a balanced diet for teenagers. Assessment activity 7.1	**2A.P2** Develop a diet and exercise plan based on level and type of exercise and appropriate nutritional balance, to promote healthy living for an individual. Assessment activity 7.1		
1A.3 Identify measures taken to improve the health of the population. Assessment activity 7.2	**2A.P3** Describe the ways in which health improvement measures are intended to improve the health of the population. Assessment activity 7.2	**2A.M2** Analyse rates of disease in the population in relation to lifestyle choices. Assessment activity 7.2	**2A.D2** Evaluate measures taken to improve the health of the population. Assessment activity 7.2
Learning aim B: Know how preventative measures can be used to support healthy living			
1B.4 Identify the role of the immune system in defending the body. Assessment activity 7.3	**2B.P4** Describe how the immune system defends the body in relation to specific and non-specific immune responses. Assessment activity 7.3	**2B.M3** Compare the different defence mechanisms the immune system uses to protect the human body. Assessment activity 7.3	**2B.D3** Evaluate the effectiveness of human vaccination and screening programmes. Assessment activity 7.4
1B.5 Identify how a vaccine aids in defending the body. Assessment activity 7.3	**2B.P5** Describe the changes in the human body following vaccination. Assessment activity 7.3		
1B.6 Identify screening programmes. Assessment activity 7.4	**2B.P6** Describe the role of specific health screening programmes. Assessment activity 7.4	**2B.M4** Discuss the advantages and disadvantages of a specific health screening programme. Assessment activity 7.4	

Learning aim C: Investigate how some treatments are used when illness occurs

1C.7	2C.P7	2C.M5	2C.D4
Describe how antibiotics are prescribed for use. Assessment activity 7.5	Investigate the use and misuse of antibiotics, using secondary data. Assessment activity 7.5	Analyse the effectiveness of different kinds of medical treatment in health-care, using secondary data. Assessment activity 7.6	Evaluate the use of different kinds of medical treatments, justifying your opinions. Assessment activity 7.6 Assessment activity 7.9
1C.8 Identify pathogens that cannot be treated by antibiotics. Assessment activity 7.5	**2C.P8** Describe the use of anti-fungal, antiviral and analgesic treatments. Assessment activity 7.5 [part] Assessment activity 7.6 [part]		
1C.9 Identify the different blood groups. Assessment activity 7.7	**2C.P9** Explain the importance of blood group matching in blood transfusions. Assessment activity 7.7	**2C.M6** Describe organ donation and approaches used to reduce rejection. Assessment activity 7.8	**2C.D5** Evaluate the potential benefits of stem cell therapy. Assessment activity 7.9

How you will be assessed

You will have lots of opportunities to complete internally assessed tasks. There are nine assessment activities in this unit and your teachers will also set you some large assignments. Edexcel will sample your centre's assessed work and verify the grades you have been awarded.

You will be expected to show an understanding of biology in the context of factors that promote healthy living, preventative measures to support healthy living and some treatments that are used when illness occurs.

Some of the assessment tasks are in the context of various workplace scenarios, which place you, the learner, in such positions as being a dietician or working in health promotion.

Your assessment may be in the form of

- written reports of experiments you will carry out
- posters and leaflets
- articles for magazines

all of which may include diagrams or tables.

7.1 Diet and health

Get started

Make a list of all the processed foods you have eaten during the last week. Include ready meals, tinned food and fizzy drinks. Estimate how much sugar is in these foods. You may be able to find this information on the food labels on the packaging.

What is a balanced diet?

A balanced diet contains the right amount of energy you need for growth, development and activity, and the right amount of different nutrients to keep you healthy. The table below lists these nutrients and some of the foods they are found in. Some of the nutrients you need, such as water, fibre, vitamins and minerals, do not give you any energy but are essential for health.

What you need to eat depends on your age and your level of activity. If you are growing and run around a lot, you will need to eat more than someone who is not growing and sits at a desk all day, because the more active you are, the more energy you need.

Nutrient	Examples of what food you need to eat to obtain it	Why you need it
Carbohydrate	Bread, couscous, pasta (all made from flour), rice, potatoes	For energy
Fats	Vegetable oil, butter, cream, cheese	For energy; essential fatty acids are needed for growth – especially of brain and nervous system; for storage under skin and around organs for protection
Proteins	Meat, fish, eggs, beans, milk, cheese	For growth and repair; for muscle, haemoglobin, enzymes, antibodies
Vitamins	Liver, milk, eggs, fruit and vegetables	To regulate growth and development; to reduce your risk of heart attacks, strokes and cancer
Minerals	Liver, milk, eggs, fruit and vegetables, wholemeal bread or rice	Iron for haemoglobin; calcium for bones and teeth; sodium and potassium for nerves
Fibre	Fruit and vegetables, wholemeal bread	To help food pass along your gut; to reduce the risk of bowel cancer
Water	Water and drinks made with water; many foods, especially fruit and vegetables	To make new cells and body fluids, like blood; to replace water lost in urine, faeces, tears and sweat

Key term

Obesity – A medical condition in which someone is very overweight, due to an abnormal increase in the number of fat cells around internal organs or under the skin, and has a BMI over 30.

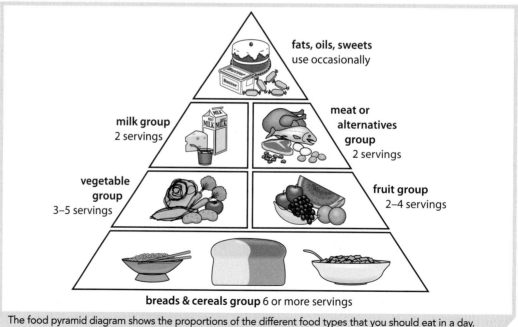

fats, oils, sweets
use occasionally

milk group
2 servings

meat or alternatives group
2 servings

vegetable group
3–5 servings

fruit group
2–4 servings

breads & cereals group 6 or more servings

The food pyramid diagram shows the proportions of the different food types that you should eat in a day.

1 Write down everything that you ate yesterday. Analyse your diet by stating which nutrients are in each food that you ate.
(a) Do you think you ate a balanced diet yesterday?
(b) Was yesterday's diet typical for you or unusual?

2 We are advised to eat *at least* five portions of fruit and vegetables each day, ideally *eight* portions. A portion is one fruit, e.g. an orange, apple, or banana, or two tablespoons of vegetables. How many portions of fruit and vegetables do you eat each day?

What happens if your diet is not balanced?

Eating too much

If you eat too much food you will take in more energy than you need for your growth and level of activity. This excess energy will be stored as fat and you will gain weight. This can lead to **obesity**. Being obese puts you at greater risk of type 2 diabetes, heart disease, strokes, arthritis and cancer.

Person	Recommended average per day	
	kilocalories	kilojoules
Adult	2500 (male) 2000 (female)	10 500 (male) 8400 (female)
Teenager (15–18)	2755 (male) 2110 (female)	11 571 (male) 8862 (female)
Baby (newborn)	690 (male) 515 (female)	2898 (male) 2163 (female)
Elderly person	1900 (male/female)	7980 (male/female)

- Too much saturated fat, such as that found in meat products, increases your risk of heart disease and stroke.
- Too much sugar may lead to type 2 diabetes, obesity and heart disease.
- Too much salt increases your risk of high blood pressure, heart disease and stroke.
- Some vitamins are toxic if you eat too much of them, for example, too much vitamin A can cause damage to the nervous system.

Eating too little

If your energy intake is not enough for your energy needs for growth, repair of body tissues, movement and **metabolism**, you will lose weight as you use stored fat to make up the energy shortfall. You may also lack some vitamins and minerals. People who are trying to lose weight by reducing the amount of food they eat may take vitamin and mineral supplements.

Some people develop eating disorders such as **anorexia**, where they eat too little food. Children or teenagers with anorexia:

- do not grow or develop properly
- do not go through puberty
- have wasted muscles
- are very thin
- feel cold
- suffer from deficiency diseases as their diet does not contain enough vitamins or minerals.

Take it further

The ancient Egyptians knew that feeding liver to a person with night blindness cured them of that condition.

In 1747, a Scottish surgeon realised that eating citrus fruits could prevent sailors from developing scurvy. Scurvy is a condition where wounds won't heal, you bruise easily and your teeth fall out due to swollen bleeding gums. During the twentieth century, scientists discovered vitamins and found out why we need to have them in our diet.

1 Which vitamin in liver cures night blindness?

2 Which vitamin in citrus fruits prevents scurvy?

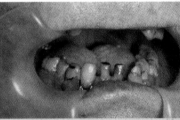

Scurvy can damage your gums and cause your teeth to fall out.

Just checking

1 Why do you need to eat fat?

2 Why do you need to eat protein?

3 From the list of foods below, which should you eat to:
(a) strengthen your bones
(b) avoid becoming anaemic
(c) avoid getting scurvy?

Milk; lamb kebabs; white bread; green peppers; chocolate biscuits.

Lesson outcome

You should be able to describe and explain the concept of a balanced diet and how dietary imbalance may lead to disorder in the human body.

Physical effects of exercise

Some exercise is essential for healthy living. It helps you grow properly, develop muscles and strengthen your bones. It also uses energy from your food because muscles need energy to contract. Taking regular exercise helps you to maintain a healthy weight.

Activity A

1 Write down everything you did each day for the last week that involved movement.
2 Write down how many times a day you sit still for more than 5 minutes at a time and also note how long each sitting period lasts.
3 Do you think you need to move about more or do you move enough?

It is also possible to over-exercise, when negative effects such as strain on your tendons, muscles and bones, your heart and your immune system can occur.

Throughout your life, exercise helps improve your heart and circulatory system – it improves your **cardiovascular** health. Exercise also improves your respiratory (breathing) system fitness, your muscles, bones and joints, and your general health. The table lists some of the benefits of exercise.

Cardiovascular system	Respiratory system	Musculoskeletal system	General health and well-being
Increases volume of blood pumped out at each beat	Increases volume of air breathed in and out at each breath (**tidal volume**)	Increases muscle size	Decreases resting blood pressure which helps reduce risk of heart disease and stroke
Lowers resting heart rate and resting pulse rate		Increases amount of glycogen (carbohydrate) and fat stored in muscles	Helps you to maintain a healthy weight
Increases size of heart muscle		Increases bone density	Improves your immune system and resistance to infection
Decreases resting blood pressure		Strengthens ligaments and tendons	Reduces risk of lower back problems
Increases blood supply to muscles		Improves flexibility, strength and coordination	Produces chemicals called **endorphins** which act on your brain to make you feel good and reduce stress

Children enjoying exercise.

Assessment activity 7.1 | 2A.P1 | 2A.P2 | 2A.M1 | 2A.D1 |

1 You are a dietician and have been asked to advise a secondary school about meals. Produce a leaflet to explain the importance of a balanced diet. Include sample menus to suggest a balanced diet for teenagers.

2 Describe how the following in your diet can affect how your body works: **(a)** not enough iron, **(b)** too much saturated fat, **(c)** too much salt, **(d)** not enough vitamin C, **(e)** not enough protein, **(f)** not enough fibre.

3 Describe the effect of exercise on the functioning of the human body.

4 You are employed as a personal health trainer. Develop a diet and exercise plan for a teenage client, taking into account their weight and level of fitness.

5 Explain how your diet and exercise plan for a teenager will affect how their body functions.

6 Evaluate your diet and exercise plan, and justify the menus and exercise activities you have chosen.

Tips

Your leaflet should be clear and concise but informative. Use colour, tables and/ or annotated diagrams to make it interesting and eye-catching. For your sample menus, make sure that you use items that are not too costly and are easy to find in shops. Try to design meals that are easy to prepare but also tasty and nutritious.

For 2A.P1, questions **2** and **3** could be completed together, perhaps as a poster, which could include an annotated diagram or table.

For 2A.P2, find out how many kilojoules per day male and female teenagers need. Bear in mind that they will still be growing and are likely to be physically active – think about their hobbies and sports and whether they walk to school or work. Make sure the diet you plan has enough energy and design a week's meals that will deliver all the nutrients your client needs. When planning the exercise programme, think about how often they are sitting still and how much exercise, such as sports or at the gym, they would need to do.

For 2A.M1, extend your answer by explaining *how* the diet and exercise plan you have developed helps your client's body to function well.

For 2A.D1, develop your answer further by evaluating your plans. Do the menus provide all the nutrients? How easily and cheaply can the meals be prepared? How easily can the exercise, including everyday activities, be carried out in school and at home?

Take it further

Recent scientific research shows that daily activity is important. If you don't have time to spend 5 hours at the gym every week, walking, playing sport, gardening, doing household chores or taking the stairs instead of the lift will all help improve your health.

A good walk every day leads to the production of an enzyme that reduces the amount of fat in your blood. The enzyme causes the fat to be stored in muscles where it is more likely to be used for energy instead of forming fatty deposits in your artery walls, so the risk of developing heart disease and stroke may be reduced.

Just checking

1 List three ways that exercise improves your:
   (a) cardiovascular system; **(b)** musculoskeletal system; **(c)** general health.
2 What effects could it have on your body if you don't do enough exercise?

Lesson outcome

You should understand how physical exercise can affect your health.

Since the year 2000:

- the percentage of overweight and obese people in the UK has risen
- the percentage of people drinking alcohol to hazardous or harmful levels has increased and the percentage of underage drinkers has doubled
- the percentage of people smoking tobacco has fallen from 30% to about 20%.

Unhealthy behaviours can lead to early death but also cost the health service large amounts of money – more than £10 billion per year for obesity-related, smoking-related and alcohol-related illnesses. In addition, they cost the wider economy even more money due to loss of productivity because people are absent from work.

Obesity

A simple way of estimating if a person is overweight is by calculating their **body mass index (BMI)**. This is their mass in kilograms divided by their height in metres squared.

The formula for calculating BMI is:

$$BMI = \frac{mass\ (kg)}{[height\ (m)]^2}$$

- A BMI of between 18 and 25 is normal/acceptable.
- A BMI of between 25 and 29.9 means a person is overweight.
- A BMI of 30+ means a person is obese, but this is not always accurate as very muscular athletes may have a large mass and hence have a high BMI even though they are fit and healthy.
- A BMI of less than 18.5 is underweight.

Mass (kg)													
Height (cm)	54	59	64	68	73	77	82	86	91	95	100	104	109	113
137	29	31	34	36	39	41	43	46	48	51	53	56	58	60
142	27	29	31	34	36	38	40	43	45	47	49	52	54	56
147	25	27	29	31	34	36	38	40	42	44	46	58	50	52
152	23	25	27	29	31	33	35	37	39	41	43	45	47	49
158	23	24	26	27	29	31	33	35	37	38	40	42	44	46
163	21	22	24	26	28	29	31	33	34	36	38	40	41	43
168	19	21	23	24	26	27	29	31	32	34	36	37	39	40
173	18	20	21	23	24	26	27	29	30	32	34	35	37	38
178	17	19	20	22	23	24	26	27	29	30	32	33	35	36
183	16	18	19	20	22	23	24	26	27	28	30	31	33	34
188	16	17	18	19	21	22	23	24	26	27	28	30	31	32
193	15	16	17	18	20	21	22	23	24	26	27	38	29	30
198	14	15	16	17	19	20	21	22	23	24	25	27	28	29
203	13	14	15	17	18	19	20	21	22	23	24	25	26	28

underweight   healthy weight   overweight   obese

A chart showing how BMI relates to weight and height.

Worked example

Calculate the BMI for a person who is 164 cm tall and who weighs 60 kg.

$$BMI = mass (kg)/[height (m)]^2$$

Step 1 Change the height to metres and then square it.

164 cm = 1.64 m

1.64 × 1.64 = 2.69

Step 2 Divide the mass by the square of the height in metres.

$$\frac{60}{2.69} = 22.3$$

A BMI of 22.3 is in the normal range.

Now calculate your own BMI. Which BMI category are you in?

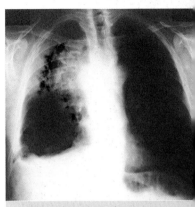

False colour X-ray showing a healthy lung and a lung with a cancerous tumour.

In the UK today, about 60% of adults and 28% of children are overweight or obese. Being obese greatly increases your risk of:

- heart disease and stroke
- high blood pressure
- type 2 diabetes
- cancer.

 ## Alcohol

Overconsumption of alcohol can cause health problems and early deaths from liver disease and cancer. Overuse also contributes to high blood pressure, heart disease and strokes, brain damage, and to levels of depression, suicide, violence, crime and dangerous driving.

 ## Smoking

Smoking tobacco can cause lung and other cancers, chronic obstructive pulmonary disease (COPD), high blood pressure, heart disease and stroke. In pregnant women, smoking can lead to low birth-weight babies, with increased risk of miscarriage and cot death. Smoking is the leading cause of preventable deaths.

 Did you know?

Adolescents often take part in risky behaviours. Scientists studying brain development, have found that during adolescence our brains are organised to focus more on the rewards of risky behaviours than on the long-term consequences. Adult brains, in contrast, focus more on the consequences.

This difference has advantages as it encourages young people to 'go out and brave the world', which can help them to be successful in life and work.

Activity A

1. Think about campaigns for a healthy lifestyle that you have seen. Did they make you change your behaviour?
   (a) If they did – why do you think they were effective?
   (b) If they didn't – why do you think they did not work?
2. Think of all the unhealthy behaviours that your friends indulge in. Why do you think they do it?

 Lesson outcome

You should appreciate the need for measures to be taken to improve the health of the nation in relation to unhealthy eating, smoking and alcohol intake.

Just checking

1. Explain why unhealthy behaviours cost the NHS a lot of money each year.
2. Explain why unhealthy behaviours are harmful to the national economy.

Government policies

The government is continually trying to reduce the harm done to the health of our population by unhealthy behaviours.

NHS health promotion campaigns such as 'Change4Life' and 'Smokefree' were set up to get people to change their lifestyles and so live longer through recommendations like:

- eat more fruit and vegetables and less processed food
- take part in more physical activity
- stop smoking
- drink alcohol sensibly – follow the guidelines.

The campaigns involve educating people to raise their awareness of the effects of their behaviour on their health. It also provides support for people trying to change their lifestyle – for example 'quit smoking' clinics at GP surgeries.

The government has increased taxation on alcohol and tobacco products, making them more expensive to buy. It has banned smoking in public places, pubs and restaurants, and there are heavy penalties for selling tobacco or alcohol to underage youngsters.

This card shows that the person is old enough to legally buy alcohol and cigarettes in the UK.

Activity A

1. Research and find out what fines shopkeepers in the UK can face if they sell tobacco to people under the age of 18 years.
2. Research and find out what fines pub landlords and licensed shopkeepers in the UK can face if they sell alcohol to anyone under 18 years old.
3. What are the laws regarding drinking alcohol in public places in the UK?
4. What are the laws regarding people under the age of 16 drinking alcohol in their homes in the UK?
5. Share your findings from questions **1** to **4** with others in your class.
6. Discuss whether the laws relating to selling tobacco and alcohol are stringent enough, and are they always able to be enforced?

Changing behaviour isn't easy

Despite the amount of information provided via campaigns like those mentioned above, getting people to change their behaviour and their habits is very difficult.

Even when people understand the problem, changing behaviour is not always easy.

- Fresh fruit, vegetables and meat are expensive compared to cheap, processed food, which contains a lot of salt, sugar and saturated fat.
- Some people do not have basic cooking skills so they rely on ready-made meals, which can be high in salt and fat.
- Many people have busy lives and say they cannot find time to exercise, or to cook meals from scratch.
- Nicotine, a chemical in tobacco, is addictive.
- Pubs and clubs have long opening hours and supermarkets sell alcohol cheaply.

| 1 large glass of 12% wine | 1 double gin and tonic | 1 double vodka | 1 bottle of 12% wine | 1 pint of 4% beer | 1 pint of 5.2% lager | 1 pint of 6% cider |

The units of alcohol in some alcoholic drinks.

- Many people know that the guidelines for alcohol consumption are a maximum of 3–4 units per day for men and 1–2 units per day for women, but do not know exactly what a unit of alcohol is.
- A binge-drinking culture has developed, with many young people thinking alcohol is essential for good socialising.

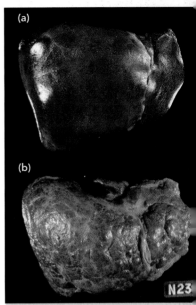

Photo showing **(a)** a normal liver and **(b)** a cirrhosis–damaged liver.

Assessment activity 7.2 | 2A.P3 | 2A.M2 | 2A.D2 |

Present your findings on the following to the rest of your class.
1 Find out about, and make a list of, measures taken by the government, NHS, schools and colleges to improve the health of the population.
2 Describe how these measures are likely to improve the health of the population.
3 Use books or the Internet to find and analyse data about rates of diseases in the population in relation to lifestyles.
4 Evaluate the measures taken to improve the health of the population.

Tips

For 2A.P3, for each measure you have listed say which illnesses it will reduce.

For 2A.M2, find out about the changes in rates of heart disease, lung cancer, COPD, liver disease and obesity. Remember that obesity is a risk factor for many other diseases.

As an example, since smoking in public places has been banned, heart attack rates have reduced by as much as 26% per year.

For 2A.D2, evaluate means find out if they worked or not and why/why not.

Think of the best way to present your findings for this assessment. Charts and graphs may be very useful. Use the Internet to find some up-to-date statistics for the UK.

Just checking

John drinks half a bottle of wine with his meal on Sunday and finishes the bottle on Monday evening. On Tuesday he calls at the pub on the way home from work and drinks 2 pints of lager. On Wednesday evening he watches a DVD at home and drinks 3 pints of cider. On Thursday he goes to the cinema and has two double gin and tonics afterwards. On Friday he meets friends and has a meal out. He has a double vodka before the meal and 2 large glasses of wine with his meal. On Saturday he goes out and drinks 3 pints of beer.

1 How many units of alcohol does John drink during the week?
2 What advice would you give John to help him improve his health?

Lesson outcomes

You should know some of the measures taken to improve the health of the nation and the effect that they have.

7.5 The body's defences against infection

Get started

Make a list of as many infectious diseases as you can think of. Now find out which microorganism causes each one. Share your ideas with others in your class.

Key terms

Pathogen – Microorganism that can invade other cells or organisms and cause disease.

Link

Remember from Tissues in Principles of Applied Science lesson 1.2 that the space in your gut is actually outside of your body. So is the space in your lungs.

Your body is continuously exposed to **microorganisms**, many of which are harmless but some of which could be **pathogens** that cause **infection**. Pathogens are:

- on the food you eat
- in the air you breathe
- on your hands after you have touched surfaces
- living on your skin
- living in your gut.

Your body uses different processes to protect you from harmful microorganisms so you don't become ill.

Physical barriers

- Inside your nose are hairs to filter out large dirt particles.
- Cells lining the gut, airways, anus and vagina produce **mucus** which traps dirt and microorganisms.
- The **cilia** lining your airways are also a physical barrier. The cilia beat to sweep the mucus up to your throat. You may cough it out or swallow it.
- Your skin makes a physical barrier that microorganisms cannot normally cross. If you break this barrier, for example a cut, microorganisms may be able to enter your body tissues and cause damage.
- When you cut yourself, your blood clots to plug the wound and prevent the entry of pathogens. If any pathogens do enter, special proteins in the blood can be activated to destroy bacteria.

Take it further

What are the symptoms of whooping cough? There is currently an increase in the number of cases of whooping cough amongst young adults in parts of the UK. Try to find out why this is happening.

Whooping cough bacteria (coloured yellow) in the trachea (windpipe). Notice the cilia, the fine hair-like structures that line the trachea.

Did you know?

Dogs have much more acid in their stomachs than humans do. This is why they can drink dirty water from puddles and not be ill.

Chemical defences

- Cells in the lining of your stomach make hydrochloric acid. This makes the pH of your stomach contents very low, pH 1–2, to kill bacteria.
- Your blood, tears, mucus and saliva contain **lysozyme** – an enzyme that can break down the cell walls of bacteria and destroy them.

Activity A

Human breast milk also contains lysozyme.
Babies fed on formula feed are more likely to suffer from diarrhoea than are breastfed babies. Why do you think this is?

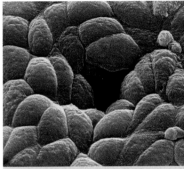

Part of stomach lining showing a gastric pit. Some cells inside the gastric pits make acid.

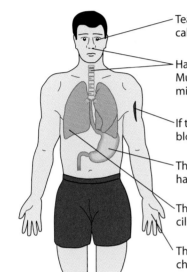

Tear glands make a liquid containing enzymes called lysozymes that kill microorganisms.

Hairs in the nose filter out dust that might carry pathogens. Mucus in the nose, throat and breathing passages traps microorganisms.

If the skin is broken, the blood clots to block entry by pathogens.

The stomach makes hydrochloric acid to kill harmful microorganisms in food.

The tubes in the lungs also produce mucus. Tiny hairs called cilia sweep out the mucus and microorganisms trapped in it.

The skin forms a protective barrier. Sweat glands in skin make chemical substances that kill harmful microorganisms.

How the human body is protected from pathogens.

Common misconception

Many people think stomach ulcers are caused by stress or poor diet. In fact they are caused by bacteria that can survive the acid conditions of the stomach. A scientist in Australia discovered this and deliberately infected himself with the bacteria and got an ulcer. He got the Nobel Prize for his work. Thanks to his work, nowadays people can be cured of stomach ulcers with antibiotics.

Activity B

Bacteria can protect you from pathogens. On your skin and in your gut you have resident bacteria that are necessary for your survival and they also help exclude invading bacteria. How do you think they might do this?

Just checking

1 State three ways that bacteria and viruses can enter your body.

2 Where is lysozyme found and what does it do?

3 Name another chemical barrier to prevent microorganisms entering your body.

Lesson outcome

You should appreciate that the human body's defences include physical and chemical barriers to prevent entry of microorganisms.

7.6 The immune system

Get started

Have you ever been stung by a bee or wasp? Do you suffer from hay fever? Have you ever had a splinter or thorn in your finger? What symptoms did you notice? What medication did you use?

Key terms

Antibodies – Proteins produced in the blood that attack and kill pathogens such as bacteria or viruses.

Inflammation – Part of the non-specific immune response, involves swelling and pain.

Phagocytosis – The process carried out by phagocytes when they ingest a harmful particle, such as smoke in the lungs, or a pathogen, such as a bacterium or virus.

Link

Read lesson 7.5 for more information on physical barriers and chemical defences.

Did you know?

A macrophage can ingest up to 100 bacteria in a second. Pus at a wound is mainly dead macrophages that have eaten themselves to death.

Non-specific immune response

White blood cells are made in your bone marrow and carried in the blood to your tissues. White blood cells called macrophages are also phagocytes (*phago* means eating and *cyte* means cell) because they detect invading pathogens and ingest (eat) them. Then they produce enzymes and chemicals to destroy the pathogens they have ingested.

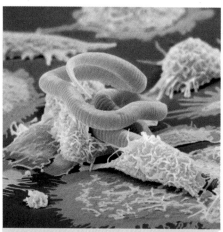

Macrophages attacking a worm parasite.

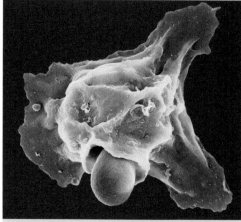

Neutrophil, another type of phagocytic white blood cell, ingesting thrush-causing fungus.

When you cut yourself and dirt and bacteria get into the wound, you may have noticed that the area becomes swollen and hot. This is due to **inflammation**, which involves:

- damaged cells releasing a chemical called **histamine**
- histamine acting as a signal to a type of phagocyte called a **macrophage** which contains digestive enzymes
- the macrophages surrounding and digesting the bacteria in a process called **phagocytosis**.

Inflammation and phagocytosis are your body's first line of defence if pathogens get into your tissues.

Physical barriers and chemical defences are also part of the non-specific immune system.

Activity A

1 Make a flow diagram to show what happens during inflammation after a splinter enters your finger.
2 When you have been bitten by an insect you can rub on some cream called an **antihistamine**.
   (a) How do you think antihistamines can help reduce inflammation?
   (b) Do you think it is a good idea to reduce inflammation?

Specific immune response

If any pathogens are not killed by the macrophages then other white blood cells called **lymphocytes** take over. There are two types of lymphocytes – B cells and T cells.

- One type of T lymphocyte kills virus-infected cells and cells that may have become cancerous.
- B lymphocytes produce **antibodies** to kill pathogens or destroy the toxins they produce.

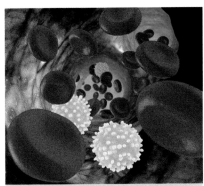

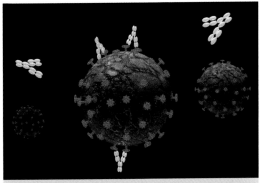

Two lymphocytes with some red blood cells. Notice that lymphocytes are about the same size as red blood cells. They consist mainly of a nucleus, with very little cytoplasm.

Computer artwork showing antibodies surrounding a virus.

Activity B

Make a large, annotated flow diagram to show the stages involved when your body's immune system carries out a specific immune response to the flu virus.

Pathogens have chemicals called **antigens** on their surface. Each type of pathogen has particular antigens on its surface. We also have antigens on our cell surface membranes. Your immune system can tell the difference between your antigens and those of an invading pathogen.

One of your B lymphocytes recognises the antigens on the pathogen and multiplies into lots of identical B lymphocytes which make many antibodies. The antibodies can then latch on to the pathogen's antigens and coat the pathogen. This makes it easier for macrophages to ingest and destroy the pathogen. It also stops pathogens entering your cells. This is called a **specific immune response**.

Natural acquired immunity

If you are infected by a pathogen, such as a virus or bacteria, and your immune system overcomes it, after you have recovered you are immune to that infection and don't become ill if you come into contact with the pathogen again.

This happens because when a B lymphocyte attacks a pathogen it makes **memory cells**. These memory cells remain in your body for many years. If your body is attacked later by the same invading pathogen:
- the memory cells release antibodies very quickly
- the antibodies destroy the pathogen before you become ill.

This is called **natural acquired immunity**.

Common misconception

Some people confuse antigens and pathogens but they are not the same thing. A pathogen is an organism that gets inside your body and makes you ill, for example bacteria and viruses. These organisms have special molecules on their surfaces, called antigens. Your immune system can recognise the pathogen's antigens because they are different from the antigens on your own cells.

Just checking

Put these events that take place when your body mounts an immune response in the correct sequence.
A This B lymphocyte is activated and divides into many identical B lymphocyte cells.
B Macrophages display the bacterial antigens on their cell surface.
C Some of these newly-made B lymphocytes make lots of antibodies; the rest of them become memory cells.
D Macrophages ingest and digest the invading bacteria.
E Bacteria enter your blood through a cut in your skin.
F The display of antigens alerts a B lymphocyte that has antibodies that fit these antigens.

Lesson outcomes

You should understand that if microorganisms enter the body then the first line of defence involves inflammation and phagocytosis, which are non-specific responses, and understand the specific immune response and how this gives you natural acquired immunity.

7.7 Vaccination

Get started

Make a list of the diseases you have been vaccinated against. Do you know when you were vaccinated for each one, and did you need a booster? Which diseases might you need to be vaccinated against when you go travelling abroad?

Key term

Vaccination – A dead or weakened pathogen, or its antigens, introduced into the body to make the body carry out a specific immune response and protect it from the disease the pathogen causes.

Common misconception

Many people think infectious diseases like measles, mumps and rubella are trivial. This is not true as, apart from causing death in rare cases, measles can cause brain damage, rubella during pregnancy leads to defects of the fetus and mumps can cause deafness or sterility.

Activity A

Find out about the vaccination schedule in the UK for children. In small groups, discuss whether the advantages of vaccinations outweigh the possible risks.

Science researchers have found out how to make you immune to an infection without suffering the disease first – they make **vaccines**.

When you are vaccinated (injected with the vaccine), your body responds by mounting an immune response. The vaccine contains either:

- a dead or weakened pathogen, or
- the pathogen's antigens.

A B cell is activated so it divides and makes antibodies and memory cells. The memory cells stay in your body and make you immune to the disease.

There are potential disadvantages, as well as advantages, to vaccination.

Advantages

- If everyone in a community is vaccinated then the pathogen cannot infect anyone and cannot spread. This is called **herd immunity**. However, it depends on everyone being vaccinated during childhood.

- Many infectious diseases can cause death or may leave survivors disabled in some way. Treatment also costs the NHS money. Prevention is better than cure.

- In 1967, the World Health Organization (WHO) started its smallpox eradication programme. **Vaccination** was used to prevent the spread of smallpox and it was eventually eradicated by 1980. The WHO is now on the verge of eradicating polio, also by using vaccination.

Disadvantages

- Some people may suffer side effects or a reaction to the vaccine.

- If the vaccine consists of a weakened but live pathogen, it could make some people ill with the disease, but this is very rare.

- Some people object to being vaccinated on religious grounds, as they consider it an invasive procedure because foreign material is injected into the body.

- The whooping cough vaccine has been linked to causing brain damage in a few children, so those with a family history of epilepsy are not vaccinated.

- There have also been scares, and a suggestion that the combined measles, mumps and rubella (MMR) vaccine could cause autism. This piece of research has not been supported, and the doctor who made the claims has been struck off the medical register. However, some parents have chosen to not have their children vaccinated and as a result there have been recent outbreaks of measles.

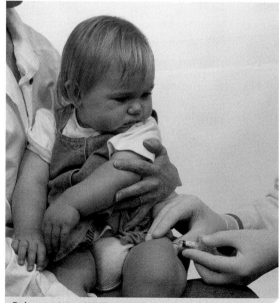

Baby receiving MMR vaccine.

Case study

Jensen works at an institution that specialises in research into tropical diseases. He used to be a nurse but after taking a postgraduate qualification in epidemiology and statistics he began working as an epidemiologist. Jensen undertakes research into the factors that affect the incidence of diseases, and he evaluates the effectiveness of preventative programmes. He is soon going abroad to an African country, Malawi, to supervise a research programme to find out about the incidence of human immunodeficiency virus (HIV) and tuberculosis (TB) infection in the population, and to find out how effective the vaccine against TB is for adults. There is a link between HIV and TB. Most people are exposed to the bacteria that cause TB, but if your immune system is healthy, macrophages in your lungs ingest the bacteria and keep them imprisoned for years. However, if someone's immune system is weakened, the TB bacteria can 'break out' from the macrophages, infect the lungs and make the person ill.

1 Why do you think many people who are HIV positive – infected with the virus but have not yet developed AIDS – develop TB?

Assessment activity 7.3

2B.P4 | 2B.P5 | 2B.M3

As a health promotion specialist, you have been asked to provide posters for a GP surgery to inform young mothers about vaccination. Your poster should include the following.

1 Draw a large diagram showing the body outline. Label it to show how the non-specific immune system prevents entry of pathogens.

2 Annotate your diagram to describe how the non-specific immune system protects us from infection. State also how the specific immune system defends us against infection.

3 Draw a flow diagram to describe the specific immune response.

4 Write a paragraph or draw a flow diagram to show how a vaccine helps to defend us from infection.

5 Write a description of what happens in the human body after a vaccination.

6 Make a table to compare non-specific immunity with acquired immunity.

Just checking

1 What is herd immunity?

2 List two advantages and two disadvantages of vaccination. On balance, do you think the advantages outweigh the disadvantages, or not?

Tips

For the non-specific immune system, think about the roles of skin, stomach, nose, airways, sweat, tears, macrophages and resident bacteria.

For the specific immune system, think about lymphocytes and antibodies.

For 2B.P4, use the labels on your diagram to provide more information. Keep the annotations clear and concise.

Constructing a flow diagram helps you to understand the process. Use arrows to show the sequence of events.

For 2B.P5, you must describe how being vaccinated stimulates an immune response and leads to the production of memory cells. Explain the role of the memory cells if this pathogen enters the body again.

You will need to carefully plan the table for 2B.M3, to compare how the different defence mechanisms protect you. Think about which are non-specific and which are specific (acquired). Explain what is meant by specific and non-specific. Think about the types of cells and tissues involved.

Lesson outcomes

You should understand how vaccination produces artificial acquired immunity, and be able to discuss the advantages and disadvantages of vaccination.

Get started

Why do you think it is important that pregnant women attend antenatal classes and are screened for various health problems? What health behaviours should people adopt when they are thinking about starting a family? Why do you think they should adopt these behaviours *before* they become pregnant?

Key terms

Health screening – Testing of a specific group of the population, before they show any symptoms of a disease.

Antenatal – Before birth.

Antenatal clinic.

- **Health screening** aims to detect early stages of a disease so treatment can begin sooner.

- There are many routine **health screening** tests offered to pregnant women. Most will have normal results, but for those who do not there are further **antenatal** tests and treatment available to prevent serious problems.

The pregnant woman's doctor or midwife will ask her questions about her general health, family history, social history and previous pregnancies (how many and were there any complications). This session takes place early in the pregnancy. If the answers indicate that there may be a risk of complications, such as **pre-eclampsia**, a dangerous complication of pregnancy, or diabetes, then the healthcare team can give special attention to these issues.

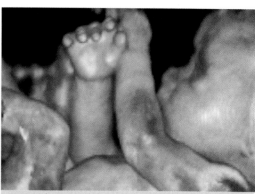

A 3D ultrasound image of a fetus.

Activity A

Why do you think pregnant women are screened to see if they are immune to rubella?

Down syndrome

Down syndrome is a genetic abnormality. If something goes wrong when cells divide to make eggs or sperm, then a sperm nucleus or egg nucleus may have an extra copy of **chromosome** 21, giving it 24 instead of 23 chromosomes. If that egg or sperm joins with a normal sperm or egg at fertilisation, the resulting **zygote** will have 47 chromosomes. It has three copies of chromosome 21.

This extra chromosome causes problems with the way the baby's body and brain develop. However, the symptoms of Down syndrome vary, ranging from mild in some cases to severe in others. The table shows the range of symptoms seen at birth and later in life.

Common physical signs of Down syndrome at birth	Other problems seen in adults
Floppy body due to decreased muscle tone	Chronic constipation
Flattened nose	Vomiting
Excess skin at the nape (back) of the neck	Hearing problems
Wide short hands with a single palm crease and short fingers	Eye cataracts
Small ears and mouth, and upward slanting eyes	Heart defects
White spots on the iris of the eye	Risk of dislocated hips
Separated joints between the bones of the skull	Underactive thyroid gland
	Increased risk of developing leukaemia

Screening for Down syndrome

Between 10 and 20 weeks into their pregnancy, women are offered a blood test. This looks for a particular protein in the blood and will indicate if there is an increased risk of the baby having Down syndrome. Older mothers are at higher risk of having a baby affected by Down syndrome.

The blood test cannot say definitely whether the baby has Down syndrome, so the parents have to decide whether or not to have a diagnostic test to find out for sure. There are two types of diagnostic test – amniocentesis and chorionic villus sampling. Both are invasive and could cause miscarriage, so there is a risk of losing a healthy fetus.

Adolescent with Down syndrome.

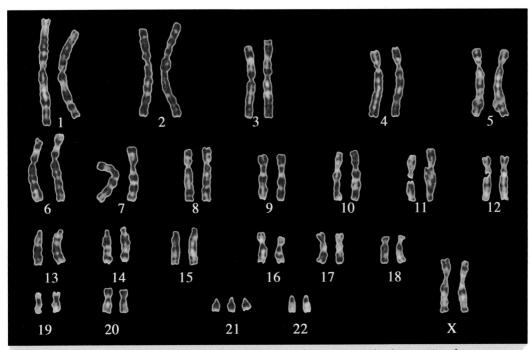

Karyotype showing chromosomes of a female with Down syndrome. Notice the three copies of chromosome 21. There are two X chromosomes showing that this is a female.

If the **karyotype** test shows that the baby has Down syndrome, the midwife will discuss the options with the parents. Whatever their decision, the parents will need counselling and support from their healthcare professionals.

Activity B

Think of one advantage and one disadvantage of antenatal screening for Down syndrome.

Just checking

1 List three physical signs in a baby with Down syndrome.
2 List three problems in adults with Down syndrome.
3 Why are teenage girls vaccinated against rubella?

Lesson outcome

You should understand the advantages and disadvantages of antenatal screening programmes.

Get started

In pairs, discuss what you know about inherited conditions such as sickle cell disease and cystic fibrosis. Do you know anyone who suffers from these conditions?

Key term

Phenylketonuria – A rare inherited disorder where sufferers cannot process phenylalanine; causes irreversible brain damage if not treated.

Link

Read Principles of Applied Science lessons 1.5 and 1.6 to remind yourself about inheritance patterns.

The NHS UK Newborn Screening Committee recommends that a blood test is offered for all newborn babies to screen for certain inherited conditions including:

- **phenylketonuria** (PKU)
- sickle cell disorders
- cystic fibrosis.

All these conditions are rare but serious, so screening all newborns allows early treatment to prevent disability or death. The test involves analysing blood spots taken from the baby's heel at about 5 days of age.

Phenylketonuria (PKU)

PKU is a rare inherited disorder with a recessive inheritance pattern. Babies with the condition inherit two faulty alleles of the gene that would normally make an enzyme called phenylalanine hydroxylase. Without this enzyme, their bodies cannot use an amino acid, **phenylalanine**, found in many protein foods. Normally this amino acid is used to make melanin, the pigment in your hair and skin. Children with PKU cannot do this, so:

- they are very fair
- the phenylalanine builds up in their bodies and causes irreversible brain damage.

The brain damage leads to severe learning difficulties and behaviour problems, meaning that children with PKU need constant care. However, if the condition is detected within the first 2 weeks of their life, their diet can be regulated.

- Babies found to have PKU are given a special milk substitute that contains very little phenylalanine.
- After weaning, the babies' diet is carefully regulated to contain limited amounts of phenylalanine.

This treatment prevents the brain damage and manages the condition, so these children can be normal and healthy.

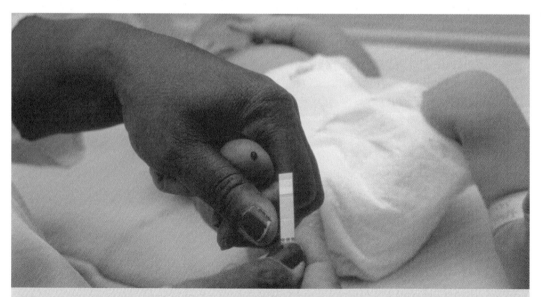

This newborn baby is having blood taken from the heel. The Guthrie test screens for PKU. The baby's blood is smeared onto filter paper and cultured with bacteria. Phenylalanine promotes the growth of these bacteria. A high level of bacterial growth indicates a high level of phenylalanine in the baby's blood.

Cystic fibrosis

Cystic fibrosis (CF) is the most common genetic inherited condition in the UK. One in 25 people carry the faulty allele. This disease was first recognised in the 1930s. At that time 70% of babies with CF died within a year. Now, with a combination of early diagnosis and swift treatment, CF patients can live into their 40s. There has been a test available to detect whether a baby has CF since the 1980s. However, babies with CF weren't tested and diagnosed until they became unwell and by that time some damage would have occurred in their lungs and digestive system and their growth rate would have been impaired. In 2007 a national screening programme for all newborn babies was introduced. This is done when the baby is 1 week old.

Potential disadvantages of screening

There are some concerns over the use of newborn screening programmes. New techniques mean that more diseases can be detected, but even though the disease can be detected, sometimes there is still no cure. If nothing can be done to cure the disease, is early detection helpful to the baby or parents? Some healthcare providers are concerned that if newborn screening is expanded there will not be effective follow-up treatment available, and that false positive results could cause parents to worry unnecessarily.

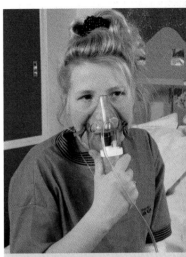

Cystic fibrosis patient using a nebuliser. Inhaling its fine spray helps to loosen mucus in the lungs.

Case study

Anya is a senior midwife, who has been practising for 20 years. She is part of a multidisciplinary team and sometimes works in a hospital, and sometimes in the community, seeing her patients at a GP surgery. She liaises with the ultrasound sonographer, obstetrician, genetic counsellor, health visitor, and sometimes with social workers.

Anya's job is to support the women, their partners and the newborns in her care from the time the pregnancy is confirmed until a few weeks after the birth of the baby. After that, a health visitor takes over the care of the family. Anya needs good listening, communication and counselling skills. She also needs to be kind and caring. She needs to make sure that the mother is offered all the routine screening tests at the correct time, and must explain the results to her. Anya gives advice and counselling before and after the screening tests, and offers support if the mother suffers a miscarriage, if the baby is stillborn or has an abnormality, or if the mother has a termination. She supervises and assists at the birth, monitoring the condition of the fetus and managing the mother's labour pain. Anya also gives support and advice about how to care for the baby, including breastfeeding and bathing. As well as good communication skills, Anya needs a wide knowledge base and has had extensive specialised training. She also helps train and supervise junior midwives.

1 Under what circumstances do you think Anya would liaise with a social worker?

2 Under what circumstances do you think Anya would liaise with an obstetrician (doctor who specialises in childbirth)?

Activity A

Why do you think screening all newborn babies for CF is advantageous to the patients and to the NHS?

Just checking

1 Why are children with PKU very fair?

2 Explain why babies are screened for PKU during their first week of life.

3 What is the treatment plan for babies with PKU?

4 What are the potential disadvantages of screening newborn babies?

Lesson outcome

You should understand the advantages and disadvantages of screening programmes for the newborn.

Key term

Cancer – A group of diseases caused by uncontrolled cell division leading to the formation of tumours or growths.

In the UK, there are NHS screening programmes for breast, cervical and colon **cancer**. There is no national screening programme for prostate cancer.

Cervical cancer

Two types of human papilloma virus – HPV 16 and HPV 18 – cause 70% of cervical cancer cases. HPV types 6 and 11 cause genital warts. Worldwide, HPV is the most common form of sexually transmitted disease. Treatment is painful and costly, and if treatment is delayed, cervical cancer may be fatal.

There is a vaccine available to protect females from infection with these viruses. The NHS offers it to girls aged between 12 and 18. It is possible to pay for a vaccination course but it can cost up to £400. It can only be offered to girls and women aged between 9 and 26 as this was the age group that the clinical trials of the vaccine were carried out on.

At present, the NHS offers screening for cervical cancer to women over the age of 25. However, some women have been found to be in the advanced stages of this cancer by the time they have had a screening test.

Activity A

In small groups, discuss:

1 whether the age at which screening for cervical cancer begins should be lowered
2 whether the vaccine for HPV should be offered to a wider range of women.

Breast cancer

The UK NHS screening programme for breast cancer began in 1988. Women aged between 50 and 70 years were invited to have an X-ray of their breasts called a **mammogram**.

This age group was chosen because breast cancer is rare in women under the age of 25 and uncommon in women under 35 years old, and 80% of breast cancer cases occur in post-menopausal women. After the **menopause**, the breast tissue becomes more fatty and a tumour is more easily seen on a mammogram.

However, breast cancer screening is now going to be extended to women aged between 47 and 73 years. Anyone who has a family history of breast cancer or who can feel a lump in the breast can ask their GP to arrange a mammogram.

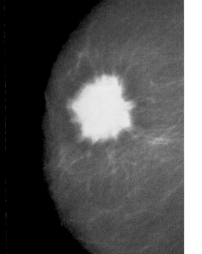

Mammogram showing a tumour in breast tissue.

Did you know?

Men can also get breast cancer. In 2008, in the UK, 341 men were diagnosed with breast cancer.

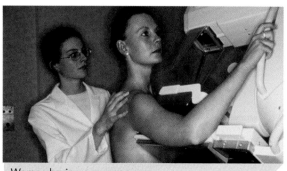

Woman having a mammogram.

Prostate cancer

Although there is no national screening programme for prostate cancer, there is an *informed choice programme* of risk management for prostate cancer. Men who are worried that they may have prostate cancer can talk to their GPs and ask for a blood test called a PSA test.

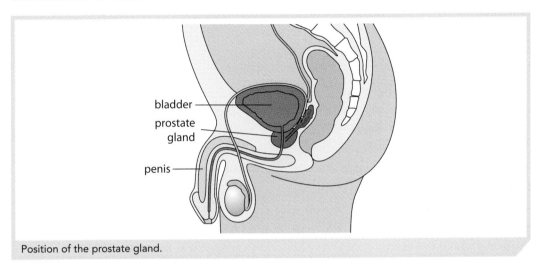

Position of the prostate gland.

- The PSA test detects a protein, called prostate-specific antigen (PSA), that is made in the **prostate gland** and leaks into the blood.
- If levels are higher than usual this could indicate cancer.
- However, two-thirds of men with raised PSA levels do not have cancer.
- And some men with prostate cancer do not have raised PSA levels.
- Some prostate cancers grow very slowly and do not lead to early death. The PSA test cannot tell slow-growing from fast-growing (harmful) cancers.

If the PSA test is positive, the man has a biopsy, which means that tissue is taken from the prostate gland and tested. If he has cancer he can have surgery and/or radiotherapy. If he does not have cancer he will have suffered an invasive and painful procedure and been very worried.

Experts disagree as to how useful the test is. Many doctors think screening the male population for prostate cancer would do more harm than good.

Activity B

In small groups, discuss whether or not you think it would be useful to have a national screening programme for prostrate cancer.

Just checking

1 One reason for vaccinating girls against HPV when they are young is because the vaccine is more effective then. Explain why it is important to protect girls from HPV infection.

2 Do you think boys should also be vaccinated against HPV? Why/why not?

Take it further

Researchers have recently discovered a new protein that only cancerous cells in the prostate make. This could form the basis of a much better test for prostate cancer.

Lesson outcome

You should understand the advantages and disadvantages of screening programmes for the detection of cancer.

7.11 Adult vascular screening programmes

Get started

Have you or any of your friends or relatives had tests to measure blood pressure and blood cholesterol levels? Did the doctor explain why these tests are done? Do you know anyone who has had a scan to assess the condition of their arteries?

Key terms

Atherosclerosis – Disease caused by build up of fatty deposits on the walls of arteries.

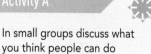

Magnified view of a slice through an artery showing fatty deposits.

Activity A

In small groups discuss what you think people can do to reduce their chances of developing atherosclerosis and so reduce their risk of having a heart attack or stroke.

Heart disease affects over 4 million people in the UK and is responsible for 20% of all hospital admissions. Most heart disease is caused by **atherosclerosis**. Atherosclerosis is caused by a build up of fatty deposits on the artery walls which causes the arteries to harden. Atherosclerosis increases the risk of heart disease, stroke and aneurysm (burst artery).

Factors that increase your chances of suffering from atherosclerosis include:

- increasing age
- having diabetes
- obesity
- lack of exercise
- high blood pressure
- high blood cholesterol
- smoking
- high alcohol intake.

Although many GPs do opportunistic screening and test patients' blood pressure and blood cholesterol levels, only about 20% of people at risk are tested in this way. Opportunistic screening means screening that takes place when the opportunity presents itself – for example, when a patient has visited the doctor for a different reason, the doctor takes the opportunity to test blood pressure and cholesterol levels at the same time.

Since June 2009, the NHS has been introducing a proactive screening programme to detect **vascular** disease in people aged 40–74 years. Proactive screening means screening where the doctor identifies patients who may be at risk and calls them in for a test before the patient has any signs of the disease. By 2013, the proactive screening programme should be available nationwide. The screening involves a scan of the abdomen to see if the aorta is widened and at risk of bursting (an aortic aneurysm). Each year, 6000 people have an aortic aneurysm and 80% die from it. Early detection of those at risk, followed by seeing a vascular surgeon, will reduce this by half.

Case study

Rafael works for a private screening company. The company travels to different towns and people can pay to have an abdominal scan to see if they are at risk of an aortic aneurysm. They can also have a scan of the arteries in their neck to see if they are at risk of having a stroke. Anyone found to be at risk of a stroke or aortic aneurysm can then see their GP and get referred to a vascular surgeon, on the NHS. This is a good example of private companies working with the NHS, but it means only those who can afford the scan are screened. Now the NHS is introducing a screening programme, which will be better, as more people at risk will get screened and deaths will be prevented. Rafael will still have a job because some people will pay to have scans more often than will be offered by the NHS.

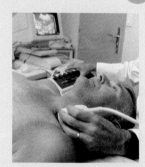

1 What are the advantages of having a scan of the abdomen and neck as described above?
2 How do you think the ultrasonographer (the person interpreting the ultrasound scan) tells if a person is at risk of an aortic aneurysm?
3 What is deposited in the walls of the neck arteries that puts someone at risk of a stroke?

Assessment activity 7.4
| 2B.P6 | 2B.M4 | 2B.D3

Use the knowledge you have gained about vaccination and health screening programmes, for different age groups in the population, to complete the following tasks.

1 Make a list or table of the screening programmes available to people. You may group the tests: for example, antenatal, newborn, adult for cancer, adult for atherosclerosis.

2 Describe the importance of the following screening programmes: breast cancer; Down syndrome; PKU; atherosclerosis. Explain why the PSA test is not described as a screening programme.

3 Choose ONE health screening programme and write a report discussing its advantages and disadvantages.

4 (a) Research, using books or the Internet, to find some statistics about how the incidence of an infectious disease has reduced following a vaccination programme. Evaluate how effective the vaccination programme has been – in terms of how many people need to have a vaccine to give herd immunity and how it has reduced the incidence of this infection. Consider also any problems caused by the vaccine.

   (b) Research, using books or the Internet, to find out how a specific screening programme has helped to reduce the harm caused by the disease it screens for. Evaluate how effective the programme has been. Has it caused any problems?

   (c) There is no screening programme for prostate cancer but there is a test available. List the reasons why you think the NHS has not introduced a screening programme for prostate cancer.

Tips

For 2B.P6, you need to explain why each screening test is important.

For 2B.M4, you should do some extra research on your chosen programme using the Internet. Do not discuss prostate cancer as there is no screening programme.

For 2B.D3, make sure you cover at least three of the screening programmes described in question **2**. Remember to include advantages and disadvantages. When evaluating, think of how effective the programme is – do the advantages outweigh any disadvantages?

Just checking

1 What is atherosclerosis?

2 List six factors that increase someone's chances of developing atherosclerosis.

3 What is opportunistic screening?

4 Each year in the UK, 6000 people have an aortic aneurysm and 80% of those die. How many people die each year in the UK from an aortic aneurysm?

Activity B

People with fatty deposits in their arteries are at greater risk of having a stroke.

How would it benefit the NHS to screen people for fatty deposits by scanning the arteries in the neck?

Lesson outcome

You should be able to discuss advantages and disadvantages of vascular screening programmes.

7.12 Antibiotics

Get started

In small groups, share information about antibiotics you have taken. What did you take them for? Did they work? Did you finish the course? Did they have any side effects?

Key terms

Antibiotics – Drugs that can destroy or inhibit the growth of microorganisms such as bacteria and some fungi, but not viruses.

Link

Look at Principles of Applied Science lesson 4.14 for more information about antibiotics and how bacteria become resistant to antibiotics.

Activity A

Claudia has flu and has felt unwell for a week. When she sees her doctor she asks for some antibiotics. What do you think the doctor will say to Claudia?

Antibiotics are drugs that are used to treat many bacterial infections such as TB, pneumonia and sepsis (blood poisoning). They work by killing or slowing the growth of the bacteria causing the infection.

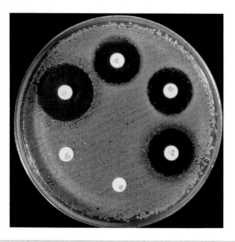

Sensitivity test: *Escherichia coli* (*E.coli*) bacteria are being grown on an agar plate. Each disc is impregnated with a different antibiotic. The one with the greatest zone of inhibition (area where bacteria do not grow) around it is the most effective against this type of bacterium.

Antibiotics, such as penicillin, that only affect bacteria are called **antibacterials**.

▶ Antibiotic resistance

Superbugs, such as the MRSA (meticillin-resistant *Staphylococcus aureus*) bacterium, are resistant to the antibiotics that are used to try to kill them or stop them reproducing. Resistant bacteria are not killed by the antibiotic but are able to survive and reproduce.

In the UK, you can only get antibiotics by prescription from a doctor or nurse practitioner. They will not prescribe antibiotics unless they are necessary, and you will be told to take the whole course, even when you begin to feel better. This is to try and reduce the occurrence of bacteria that are resistant to antibiotics.

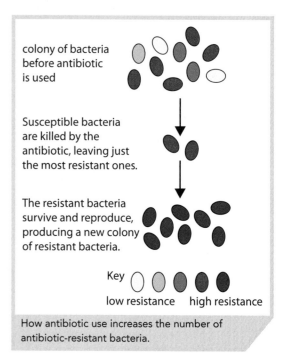

colony of bacteria before antibiotic is used

Susceptible bacteria are killed by the antibiotic, leaving just the most resistant ones.

The resistant bacteria survive and reproduce, producing a new colony of resistant bacteria.

Key   low resistance   high resistance

How antibiotic use increases the number of antibiotic-resistant bacteria.

Resistance happens by natural selection. One bacterium infecting you has a mutation in its DNA. This makes it able to resist the effects of the antibiotic. It survives, but all the other bacteria infecting you die. This survivor can now flourish as it has no competition. Bacteria reproduce by dividing into two so they pass the mutation for antibiotic resistance to all their descendants.

Other problems with antibiotics

When you take a course of antibiotics you will also kill or slow the growth of your resident bacteria. These resident or 'good' bacteria help keep down the numbers of pathogens in your body. If you kill too many of your gut bacteria, it can allow a bacterium called *Clostridium difficile* (*C. diff*) to grow in your gut. This bacterium causes severe diarrhoea and kills many hospital patients each year.

If you kill too many of the bacteria in your mouth or in the vagina this allows the thrush fungus (*Candida*) to flourish.

Some antibiotics make you feel sick if you drink alcohol while you are taking them.

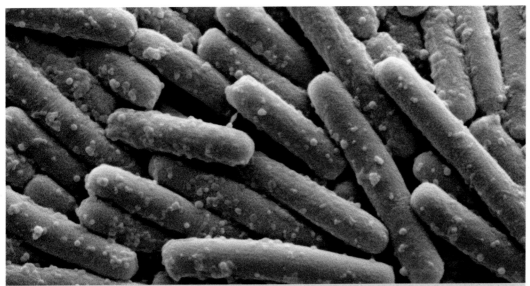

Clostridium difficile bacteria live in your gut and usually do no harm. However, if you are given antibiotics that kill many of your other gut bacteria, *C. diff* can flourish as they are resistant to many antibiotics. They can then cause severe gut infections. *C. diff* is one of the most common hospital-acquired infections.

Just checking

Say whether each of the following is true or false.
1. Antibiotics do not kill fungi.
2. Bacteria always respond to antibiotics by becoming resistant to them.
3. The bacteria that live on your skin and in your gut are called 'resident bacteria'.
4. *Clostridium difficile* bacteria kill many hospital patients each year.
5. MRSA is described as a superbug because it is resistant to many antibiotics, including meticillin.
6. Antibiotics can treat diseases caused by viruses.
7. You should always complete your course of antibiotics, unless you have nasty side effects, in which case you need a different antibiotic.
8. Bacteria become resistant to antibiotics by natural selection so this is evolution in action.
9. Antibiotics that only kill bacteria are called antibacterials.
10. MRSA is a hospital-acquired infection.

Activity B

Jeremy cut his finger and although he washed it, it has become infected with bacteria and is red, swollen and hot. His doctor has prescribed penicillin for him.

1. What instructions do you think the doctor gives Jeremy about taking the antibiotics?
2. Why is Jeremy's finger red, hot and swollen?

 Lesson outcomes

You should understand the principles, advantages and disadvantages, and use or misuse of antibiotics.

Get started

Did you know that infections such as athlete's foot, thrush, ringworm and dandruff are caused by a fungus? What are the characteristics of fungi?

Key terms

Anti-fungal – A drug that kills or slows the growth of fungi.

Antiviral – A drug used to treat infections caused by viruses, such as flu or measles.

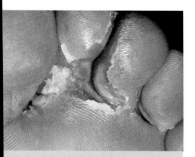

Fungal infection of the foot is the most common form of ringworm, *Tinea pedis*, infection, causing athlete's foot. If not treated early it will infect toenails, and this is more difficult to treat.

Did you know?

Tea tree oil has anti-fungal, antibacterial and antiviral properties.

Anti-fungal drugs

Diseases such as ringworm, athlete's foot, thrush, farmer's lung and cryptococcal meningitis are all caused by fungi. They are treated using **anti-fungal** drugs.

Some anti-fungal drugs interfere with the cell membrane of the pathogen and some interfere with its cell wall or its enzymes. Some drugs inhibit cell division of the fungus.

Many anti-fungal drugs can cause side effects in humans because both fungi and humans have eukaryotic cells. However, fungal cells have walls and have some enzymes that human cells do not.

For fungal infections of the skin, the anti-fungal drug is applied to your skin, causing fewer side effects. *Systemic* anti-fungal drugs that you have to take by mouth or by injection may cause liver problems. The table gives some examples of anti-fungal drugs and their uses.

Anti-fungal drug	Use
Ketaconazole	In anti-dandruff shampoo – it inhibits the yeast fungus that causes dandruff
Ciclopirox	To treat skin infections such as athlete's foot and ringworm
Zinc pyrithione and selenium sulfide	To treat dandruff
Clotrimazole and fluconazole	To treat thrush (*Candida albicans*)
Echinocandins	Interfere with cell walls of the fungus – used to treat systemic (inside the body) infections

Activity A

At swimming pools, there used to be a foot bath containing antiseptic to kill fungi. People had to walk through the foot bath before entering the pool, to prevent the spread of athlete's foot. This practice has now stopped because it had the opposite effect. Why do you think it increased the spread of athlete's foot?

In the UK, you can buy some anti-fungal drugs, for treating skin infections or thrush, at a pharmacy and of course you can buy anti-dandruff shampoo. However, for other fungal infections you need to see the doctor and get a prescription.

Antiviral drugs

Viruses do not have cell walls, membranes or any metabolism that can be interfered with to stop the infection.

Viruses enter your cells and use your cells to make copies of themselves. There are two types of **antiviral** drugs, entry inhibitors and replication inhibitors.

Entry inhibitors stop viruses entering your cells. Examples are amantadine and rimantadine which are used to combat flu.

Replication inhibitors stop viruses multiplying once inside your cells. Examples are acyclovir and zidovudine which are used to treat HIV; Relenza® and Tamiflu® which are used to treat flu; and interferon which is used to treat hepatitis B and hepatitis C.

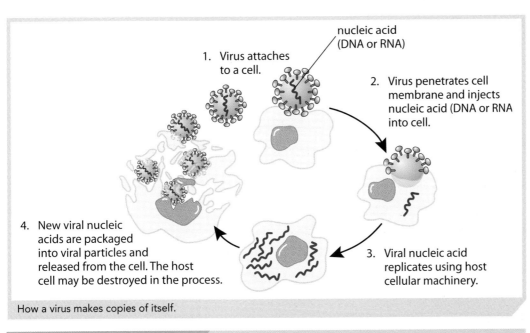

1. Virus attaches to a cell.

nucleic acid (DNA or RNA)

2. Virus penetrates cell membrane and injects nucleic acid (DNA or RNA into cell.

3. Viral nucleic acid replicates using host cellular machinery.

4. New viral nucleic acids are packaged into viral particles and released from the cell. The host cell may be destroyed in the process.

How a virus makes copies of itself.

Assessment activity 7.5

| 2C.P7 | 2C.P8 [part]

You are a trainee health technician. Your manager has asked you to demonstrate what you know about antibacterial, anti-fungal and antiviral drug treatments by completing the following tasks.

1 Make a table to show how antibiotics are prescribed for use. Include the following: **(a)** What are they used for? **(b)** How can you get them? **(c)** What advice should you follow when taking them and why?

2 Which pathogens cannot be treated by antibiotics?

3 Describe how anti-fungal drugs and antiviral drugs are used.

4 Write an article for a magazine, to be read by the general public, to explain that antibiotics are very useful and have saved many lives, but problems arise when they are misused. Your article should give advice about how to use antibiotics properly.

Research your article using the Internet, journals and magazines, and by interviewing people, to find out how antibiotics may be misused. Where can they be bought over the counter? Find out if people understand why they should take the full course and whether or not they do.

Find out about bacteria that are resistant to antibiotics and can cause serious problems (MRSA and *E. coli*).

When were antibiotics first used? Use this information when writing your article and also include a list of some of the diseases that antibiotics can treat.

Tips

Plan your article for 2C.P7 carefully. Use paragraphs and sub-sections with side headings, so it is easy to find specific pieces of information. Make a note of where you obtained your information

The information for 2C.P8 could also go into a table. Focus on anti-fungal and antiviral treatments, not analgesics.

Activity B

1 Viruses use your cells to replicate. Why do you think this makes designing safe and effective antiviral drugs difficult?

2 Some viruses mutate rapidly. Might this also cause problems for drug development?

Just checking

1 Name three anti-fungal drugs and for each one state one condition it is used to treat.

2 How do antiviral drugs known as entry inhibitors work?

3 How do antiviral drugs known as replication inhibitors work?

4 Why do many anti-fungal drugs cause side effects in humans?

Lesson outcomes

You should understand the principles, advantages and disadvantages, and use or misuse of anti-fungal and antiviral treatments.

Get started

Think of the times you have been in pain. Write down when they were, what caused them, what the pain felt like, and how bad it was on a scale of 1–10 (1 being mild and 10 being the worst imaginable). What did you do when you had these pains?

Key term

Analgesic – Drug used to reduce or relieve pain.

Did you know?

Although the brain interprets nerve impulses and allows us to feel pain in the damaged part of the body, the brain itself cannot feel pain.

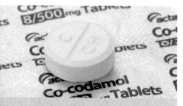

Co-codamol is a compound analgesic. It contains codeine and paracetamol and is used to treat mild to severe pain.

Opium poppies – *Papaver somniferum*. Opium obtained from the seed cases contains morphine and codeine.

Why do we feel pain?

Pain is a perception. The brain interprets sensory nerve impulses that it receives about damage to a part of the body, so that we *feel* pain in that part of the body. Pain is unpleasant and tells us that there is a problem with that body part. We can then take appropriate action and tend the wound, rest the inflamed joint or seek medical help.

- Many infections also cause pain – usually a headache and sore throat, or stomach ache.
- Some people suffer from chronic (long-term) pain that may not have an identifiable source but is nevertheless real and debilitating.
- Some people suffer severe pain after surgery or as a result of having cancer, and their pain needs to be managed.
- When people have terminal cancer, although there is no chance of recovery, they are given pain relief as part of their **palliative care**, until their death.

How can we relieve pain?

You may have used painkillers for toothache, headaches, or to reduce fever when you had a cold or flu. Painkillers do not *cure* any diseases, but they relieve some of the unpleasant symptoms. They are part of the *treatment* of an illness.

The main painkiller drugs are either **analgesics** or **non-steroidal anti-inflammatory drugs** (NSAIDs).

Analgesics

Analgesics are medicines used to reduce or eliminate pain. Some can be bought in a pharmacy, some have to be prescribed by a doctor and some are only administered under medical supervision, often in a hospital. Analgesics act on the nervous system to reduce the perception of pain. The characteristics of some analgesic medicines are given in the table.

Paracetamol	Codeine	Morphine, tramadol and pethidine
Available over the counter at a pharmacy.	Available only by prescription.	These are all opioids, and must only be given under medical supervision.
This acts on the brain to relieve pain. It is safe and can be given to children. There are few side effects but the recommended dose should not be exceeded or it could cause liver and kidney damage.	This is an **opioid** which can be taken with paracetamol. Gives stronger pain relief. Side effects include constipation.	Given after surgery or for cancer pain. Extremely effective but also addictive. Side effects include confusion, nausea, vomiting, constipation and hallucinations.

Non-steroidal anti-inflammatory drugs (NSAIDs)

NSAIDs work to reduce inflammation. By reducing swelling and relieving pressure on tissues or organs, they can also lead to pain relief. Examples are aspirin and ibuprofen.

NSAIDs inhibit enzymes involved in the inflammatory reaction. They can have side effects such as bleeding in the gut, stomach ulcers and kidney damage. Children under 16 years of age should not be given aspirin as it may cause a liver disorder called Reye's syndrome.

Activity A

1 Explain why pain is useful to us.

2 Explain the difference between analgesics and anti-inflammatories.

Case study

Tom works in a pain management clinic. He works with people who suffer from chronic pain. In these cases, drugs do not work as the body becomes tolerant. Long-term use of drugs can also lead to addiction. Tom uses cognitive behaviour therapy and teaches people that pain is very subjective. People start by keeping a pain diary, describing their pain (stabbing, burning, throbbing, etc.) and rating it on a 1–10 scale. They can learn to deal with their pain without using drugs, for example by distraction (thinking about something else that is pleasant). They may also use a TENS (transcutaneous electrical nerve stimulation) machine or acupuncture. TENS machines deliver small electrical pulses to the body from electrodes placed on the skin. Acupuncture involves trained acupuncturists inserting very fine needles (as thin as human hairs) into certain parts of the skin.

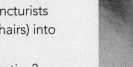

1 What is meant by saying that pain is subjective?

2 What are the advantages of managing pain without drugs?

3 Why do you think analgesics lose their effectiveness if you take them regularly over a long period of time?

Assessment activity 7.6

| 2C.P8 (part) | 2C.M5 | 2C.D4 |

1 Describe how analgesics are used in healthcare. You could make a poster or table to present this information.

2 Using the Internet, gather data to show how effective the following types of treatment are: antibiotics (antibacterial and anti-fungals), antiviral drugs, analgesics. Write a report of your findings.

3 Evaluate the different medical treatments you have discussed. For each type of treatment give one example of a condition for which it would be used. Justify your choices.

Take it further

The most effective way to reduce pain is to take an analgesic (such as paracetamol) and an anti-inflammatory (such as ibuprofen) together, instead of only an analgesic. It is even more effective if you take the drugs with a caffeine drink, such as strong coffee.

If you don't like coffee, what other sources of caffeine could you take with analgesics and anti-inflammatories to make them more effective?

Tips

For 2C.P8, you should only include analgesics. Remember those you can buy over the counter, those you get on prescription and those that would only be used in a hospital or hospice.

For 2C.M5, you are going to be using secondary data (data not collected directly by you). Make sure you include a reference list of sources for all the data you use.

Your answer for 2C.D4 could be in the form of a table. When evaluating the effectiveness of treatments, discuss the fact that analgesics only relieve the symptoms, they do not address the cause of the pain.

Lesson outcomes

You should understand the principles, advantages and disadvantages, and the use and misuse of analgesic and anti-inflammatory drugs.

Many people rely on blood transfusions to survive:

- people who have lost blood from an injury or during surgery
- people who have haemophilia; haemophilia is an inherited disease where, after an injury, blood takes a very long time to clot
- people with sickle cell disease
- some cancer patients.

Blood transfusions were first carried out during the nineteenth century. Blood was taken from one person, the **donor**, and directly given to another person, the **recipient**. But only some transfusions were successful.

In 1901, Karl Landsteiner, an Austrian doctor, discovered the ABO human blood groups. He worked out a way to check blood match between donor and recipient so that blood transfusions could become a lot safer.

Why were early transfusions unsuccessful?

On the membranes of your red blood cells there are different **antigens**. Some people have type A antigens on all their red cells, some have type B, some have none (O) and some have both A and B. This is what determines your blood group.
The liquid part of your blood tissue, in which the blood cells are suspended, is called plasma. Plasma may contain **antibodies** to the antigens but people of blood group AB have no antibodies in their plasma.

Matching recipients and donors

When a patient is given a blood transfusion, doctors use the ABO blood-grouping system to see how the antibodies of the recipient's plasma will react with the donor's red blood cell antigens. This is shown in the table.

	Group A	Group B	Group AB	Group O
Red blood cell type	A	B	AB	O
Antibodies in plasma	Anti-B	Anti-A	None	Anti-A and Anti-B
Antigens in red blood cell membrane	A antigen	B antigen	A and B antigens	None

Antigens and antibodies present in blood of different groups.

Blood group	Antigens on surface of red blood cells	Antibodies in plasma	Can donate to people of blood group	Can receive blood of group
A	type A	anti-B	A and AB	A and O
B	type B	anti-A	B and AB	B and O
AB	type A and type B	none	AB	A, B, AB and O
O	none	anti-A and anti-B	A, B, AB and O	O

People of blood group O are called **universal donors** because they can donate blood to anyone.

People of blood group AB are called **universal recipients** because they can receive blood from anyone.

Doctors also have to consider the **Rhesus factor** before giving a blood transfusion. Rhesus negative people cannot receive Rhesus positive blood.

Case study

Temi works for an organisation involved in collecting blood donations. Mobile units visit towns and places of work so people can give blood without having to travel far. Donors are given a simple pinprick test and their blood is checked to ensure that they are not anaemic. Donors also answer a questionnaire about recent travels abroad, especially outside Europe.

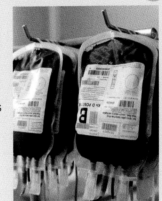

The procedure is simple and pain-free. Your body replaces the lost blood fluid within a few hours. The blood cells are replaced more slowly, but within 3 months. You can safely give blood twice a year and blood donors save many lives.

Some blood is stored whole, and some is used to make blood products such as plasma or platelets. Whole blood is stored at cold temperatures and a chemical is added to stop it clotting in the bag. Blood is also screened for pathogens and heat treated to destroy viruses.

All blood is tested and labelled with its ABO group and Rhesus factor. A barcode is used for easy identification.

1 Why do you think donors are asked about their recent travels?
2 Donors are also asked about tattoos, intravenous drug use and sexual partners. Why do you think they are asked these questions?
3 Blood is stored at cold temperatures. What do you think would happen if a patient was given a lot of blood without it being first warmed up?

Assessment activity 7.7 | 2C.P9

You are a nurse working for NHS Blood and Transplant. To help raise public awareness about blood transfusion, make a poster for a doctor's surgery or write a short article for a magazine.

You should list the different blood groups and explain the importance of blood group matching for blood transfusions.

Tip

Whether making a poster or writing an article, illustrate it with clear, labelled diagrams. Include a table showing which blood groups match for a transfusion and explain why.

Activity B

1 Explain why people of blood group A can receive group O blood.
2 Explain why people of blood group A cannot receive blood from donors with group B blood.

Take it further

There is artificial blood that does not contain red or white cells and does not have to be matched to the recipient. Artificial blood can be stored at room temperature for up to 3 years and can immediately restore the full oxygen-carrying capacity to recipients. This would take 24 hours with real blood.

Lesson outcome

You should know the principles and uses of blood grouping and blood transfusion.

7.16 Organ donation

Get started

Do you know anyone who has had an organ transplant? Do you know anyone who has donated a kidney or bone marrow? Would you be willing to have some of your organs given to someone else after you died? Discuss the reasons for your decision in pairs.

Key terms

Organ donation – Giving of an organ by a donor for transplant into a recipient. People can opt to donate organs after they die or can give some organs, such as one kidney or part of the liver, while they are living.

Rejection – Immune response of recipient that causes deterioration and death of the transplanted organ.

Link

See lesson 7.17 to find out more about rejection of transplanted organs.

Read Principles of Applied Science lessons 1.1–1.3 about cells, organs and tissues.

Who needs organ transplants?

Sometimes people suffer from organ failure. As each organ in your body is important, this will usually lead to death.

In the case of kidney failure, patients can have dialysis, where their blood is passed through a machine to have toxic waste and excess salts and water removed. This treatment involves being connected to the machine for up to 8 hours, three or four times a week. Having a functioning kidney transplanted instead greatly improves their quality of life.

When did organ transplants start?

The first successful **transplant** between humans was of the cornea (the clear, front part of the eye) in 1905. Much pioneering work was carried out in the early part of the twentieth century, particularly with skin grafts to treat patients badly burnt during the First and Second World Wars. In many cases, skin was used from another part of the patient's body so there was no problem with **rejection**.

In 1954, in the USA, a kidney was successfully transplanted between identical twins. Some landmarks in the history of organ transplants are listed in the table.

| Year | First successful transplant | |
	Organ	Country
1905	cornea	Czech Republic
1954	kidney	USA
1963	lung	USA
1966	pancreas	USA
1967	liver	USA
1967	heart	South Africa
1981	heart–lung	USA
1998	hand	France
2005	partial face	France
2010	full face	Spain
2011	double leg	Spain

Activity A

1 Your cornea does not have a blood supply. It is nourished by fluid. Why do you think the first cornea transplant was not rejected?

2 Why do you think skin transplanted from one part of a patient's body to another part is not rejected?

3 Why do you think there was no rejection of the first kidney transplanted in 1954?

Case study

Samira works for an organ donor register.

The register contains the name, sex, date of birth and address of potential donors, along with a list of the organs they wish to donate. There is no upper or lower age for joining.

By the end of 1995, more than 2 million people had joined the register, and numbers have increased 10-fold since then. The target for 2013 is to have 25 million potential donors on the register.

Samira helps organise road shows in many locations throughout England, Northern Ireland and Wales to raise awareness of the need for more organ donors and to get people to sign up. She has also helped set up a website where people can register online.

In July each year, National Transplant Week focuses on the importance of potential donors making sure their families know of their wishes. Samira works with other health organisations to help promote National Transplant Week.

1 A donor can save more than one life. List all the organs that someone could potentially donate when they die.

2 Why do you think it is important that a potential organ donor tells their family that they wish to donate organs when they die?

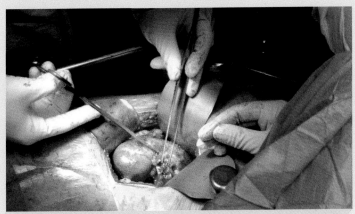
Surgeons performing a kidney transplant operation

Did you know?

There are procedures known as domino transplants. This is where a recipient with lungs that are diseased, e.g. by cystic fibrosis, receives a heart–lung transplant. These are more successful than just transplanting the lungs. The heart of the person with cystic fibrosis may be perfectly healthy and so can be given to someone in need of a heart transplant.

Activity B

Make a list of pros and cons for having a kidney transplant as opposed to needing dialysis three times a week.

Which do you think you would prefer if both your kidneys failed?

Just checking

1 Explain how a transplanted kidney may be rejected by the recipient.

2 Why do you think it is important that the blood groups of donor and recipient are matched?

3 What would happen if a kidney from a person with blood group B was transplanted into the body of a recipient with blood group O?

Lesson outcome

You should understand the principles and uses of **organ donation**.

Get started

People who have received a transplanted organ take drugs to prevent their immune system from rejecting the organ. The drugs suppress the activity of their immune system. What problems do you think these immunosuppressant drugs might cause?

Take it further

Cyclosporin was discovered by accident. Two scientists were screening soil samples looking for anti-cancer drugs. They extracted cyclosporin from a soil-living fungus. They tested it on mice but it did not work against cancer. However, they noticed that the immune systems of the mice were suppressed. Many pharmaceutical companies were not interested in producing cyclosporin, but eventually one did. It is now the most widely used immunosuppressant drug.

Activity A

Alleles of certain genes determine the shape of antigens on your cell membranes. Your immune system recognises your own antigens and can tell if a cell in your body has foreign (non-self) antigens. It treats such cells the same way as it would treat an invading pathogen.

Why do you think the transplant medical team try to find a live kidney donor amongst the recipient's close relatives?

Rejection

One major problem that can occur with organ transplants is rejection. This is when the recipient's immune system mounts a response against the foreign tissue or organ, and destroys its cells. Rejection occurs because:

- T lymphocytes recognise and attack foreign cells, because the foreign cells have antigens on their surface that are not the same as those on the recipient's cells
- B lymphocytes make antibodies against these antigens
- natural killer cells also destroy foreign cells.

There are two ways to overcome this problem of rejection:

- carry out tissue typing to find out about the antigens on the cell surface membranes of both donor and recipient, and match the tissues of donor and recipient
- **immunosuppression** – using drugs to suppress the recipient's immune response.

From the 1970s onwards, use of the drug cyclosporin has greatly increased the success rate of transplants, because it is a very effective immunosuppressant.

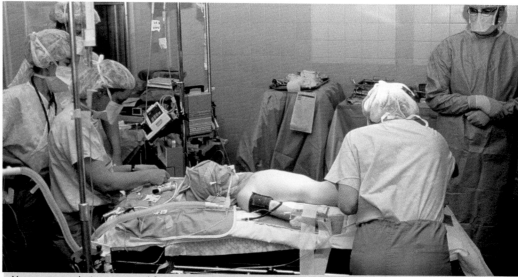

Young person being prepared for a liver transplant. As the liver can regenerate, an adult may donate part of their liver to a child. So a living donor, often a relative, may be used.

Ethical considerations

If the organ is given by a live donor, then the donor can give informed consent.

However, some organs are taken from dead people or patients diagnosed as brain dead. Some people decide that they will donate organs after their death, and they carry an organ donor card. Doctors still need to talk to the relatives, but the relatives are usually happy to comply with the deceased's wishes. Sometimes a child needs an organ transplant and adult organs are too large. This may be a sensitive issue, to ask relatives if they would donate their dead child's organs.

People may also be concerned that in some parts of the world, poor people are coerced into selling organs, such as a kidney. In some regimes, prisoners may be executed and their organs harvested.

Discussion point: Opt in or opt out

At present in the UK you need to opt in to be a donor. You make a decision and carry a card. However, today, organ transplants are fairly routine and there are more people needing organs than there are donors. Some people just do not get around to obtaining a donor card. In future we may have an 'opt out' system, like some other European countries do, where everyone is considered a potential donor, unless they carry a card to say otherwise.

Which system do you think is best? Try to think of reasons for and reasons against both systems.

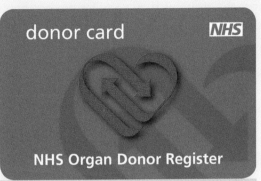

NHS donor card. This confirms that the person carrying it is willing to donate specific organs for medical use after their death.

Activity B

In small groups discuss:

1 Would you be prepared to donate one of your kidneys while still alive? If so, under what circumstances?

2 Would you be happy for your organs to be donated after you die? If not, why not?

3 Would you prefer the UK to adopt an 'opt out' system?

Today, worldwide, the most common type of organ transplant is of kidneys, followed by liver and then heart. Other organs that can be transplanted are pancreas, lungs, eyes, intestine and thymus. Tissues, such as bone, skin, heart valves, blood vessels and tendons, are also transplanted.

Assessment activity 7.8 | 2C.M6

You are a campaigner trying to increase the number of organ donors. Write a report about organ donation, to include the following.

1 Describe what is meant by 'organ donation'.

2 Indicate which organs can be donated. Mention which can be given by live donors and which by dead donors.

3 Describe how rejection of the donated organ can be reduced.

Tip

For overcoming rejection you need to refer to tissue typing and immunosuppression. Research, using textbooks and the Internet, to find out more about tissue typing and the databases for organ donation.

Just checking

1 Explain how T lymphocytes, B lymphocytes and natural killer cells are involved in rejection of an incompatible transplanted organ.

2 Explain why people who receive transplanted organs take immunosuppressant drugs.

3 What are the disadvantages of taking immunosuppressant drugs?

4 What is the most common type of organ transplant?

 Lesson outcome

You should understand the principles and uses of organ donation.

7.18 Stem cells

Get started

What do you know about stem cell research? Share your thoughts with others in your class.

Link

Find out more about specialised cells in Principles of Applied Science lesson 1.2.

Take it further

Plants have more stem cells than animals have. Plants have them at the tips of stems and roots and this is why gardeners take cuttings from plants that will grow into whole new plants. Even animals that can repair well don't do this. Lizards can grow a new (shorter) tail if they lose it, but the lost tail won't grow a new lizard.

Did you know?

In 1965, Leonard Hayflick discovered that normal differentiated cells have a limit to the number of times they can divide. This number, known as the Hayflick limit, varies among cell types and species. In humans it is 52, but may be between 40 and 60 for human fetal cells.

It is this limit that stops cells becoming cancerous and dividing uncontrollably.

What are stem cells?

Cells become specialised to perform different functions. A heart muscle cell is different from a nerve cell or a skin cell. In *specialised* cells, some genes are switched off as the proteins they code for are not needed. These specialised cells can divide, but only for a limited number of times.

Stem cells are *unspecialised* cells. All their genes are switched on and they can continue to divide for long periods even after they have been inactive. They are described as **totipotent** because they can become any kind of cell that is normally found in that species.

Sources of stem cells

You, like everyone else, started life as a stem cell – a fertilised egg. As this stem cell divided again and again, it formed a ball of unspecialised cells. Eventually, the cells became specialised and formed an embryo that grew into a baby.

There are spare embryos from **IVF treatment**. Before these embryos are discarded, scientists can obtain stem cells from them for research purposes. However, this raises some ethical concerns and some people object to the use of embryonic stem cells.

Other sources of stem cells are:

- adult bone marrow, fat tissue or blood
- babies' umbilical cords.

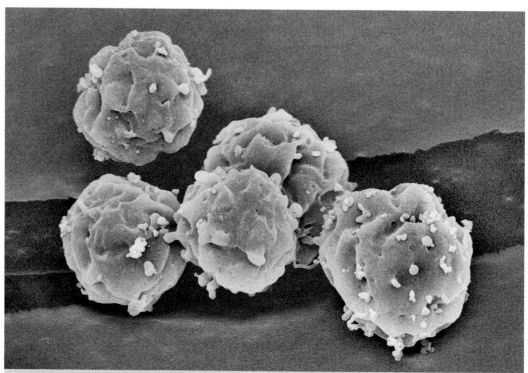

Magnified image of embryonic stem cells seen using a scanning electron microscope, with false colour added. These cells are totipotent. The type of cell they mature into depends on biochemical signals they receive before they differentiate.

Uses of stem cell therapy

Because stem cells can divide and differentiate into many cell types they have a great potential to replace diseased and damaged tissues in the body, with few side effects.

Bone marrow stem cell transplantation has been used for many years to treat leukaemia. Scientists hope that soon stem cell therapy may be used to treat:

- different forms of cancer
- wound damage
- brain damage
- diabetes
- baldness
- spinal cord injury
- **degenerative conditions** such as Alzheimer's disease and Parkinson's disease.

You have seen in lesson 7.17 that there may be a problem with the body's immune system rejecting a transplanted organ.

In **regenerative medicine**, scientists can grow stem cells taken from a person, and then stimulate the stem cells to become specialised and differentiate into the desired type of cell. They can then grow new organs. Because these organs are derived from the patient's own cells, the patient's immune system will recognise them as 'self' and not reject them.

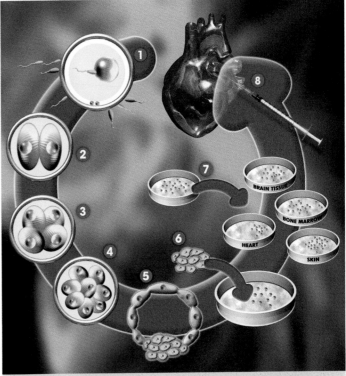

Stages involved in stem cell therapy. 1: an egg cell is fertilised by a sperm cell. 2–4: the resulting embryo divides three times over the next 3 days, producing an eight-celled structure. 5: by day 5 an early embryo has developed; the inner cells are stem cells and the outer cells will form the placenta. 6–7: the stem cells are removed and cultured into different tissue types. 8: heart muscle cells may be used to repair damaged hearts in patients.

Activity A

1 In small groups, discuss the ethical concerns of using stem cells taken from embryos. Think about the following:

- Many embryos are lost as many pregnancies naturally fail in the early stages, before the women even know they are pregnant.
- Spare embryos from IVF treatment are discarded.
- Our society allows abortion where, in 90% of cases, healthy fetuses are aborted.
- Our society also allows adults to fight in wars and lose their lives.
- An embryo is a ball of cells with no responsibilities, so does it have rights?
- Some people regard every embryo as a potential human life, to be respected.
- Stem cell research may lead to therapies that could save or improve the quality of many lives.

2 If you were working in a stem cell lab and a fire broke out, would you first save the other people in the building or a dish of embryos?

In June 2008, a team of surgeons in Spain carried out the world's first tissue-engineered whole organ transplant. The patient was a 30-year-old woman whose trachea (windpipe) had been damaged by tuberculosis. The patient's adult stem cells were taken from her bone marrow and grown in the lab. These cells were made to develop into tracheal epithelial cells and transplanted into the patient. The patient did not need to take immunosuppressant drugs, and there was no rejection.

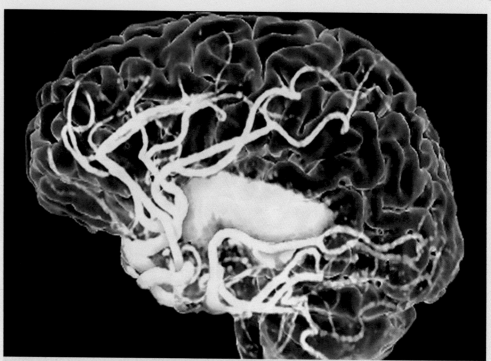

A magnetic resonance angiogram showing a human brain after a stroke. The central (yellow) region is the area damaged by the stroke.

Stroke patients in Scotland are taking part in a clinical trial. They have had neural stem cells implanted into their brains. Doctors hope the stem cells will grow and replace the patients' damaged brain cells, enabling them to regain control of speech and movement. Trials have to be carried out to find out if there are any side effects, such as causing a brain tumour or changes to a patient's personality. Early indications are that the patients are not showing any cell-related adverse effects.

1 Assuming that the new nerve cells grow and replace those damaged by the stroke, do you think these patients will have to *learn* to talk again?

Assessment activity 7.9 | 2C.D4 | 2C.D5

Make a table and *evaluate* different kinds of medical treatment, including: antibacterial and anti-fungal antibiotics, antiviral drugs, analgesics, organ transplants, blood transfusions, stem cell therapy.

Tips

For each treatment, concisely outline its use, benefits and risks/disadvantages. Use the Internet to find some up-to-date statistics about the treatments. Use these to justify your opinions/conclusions about the benefits all the treatments. Include bone marrow transplants when you consider stem cell therapy. Find out about the graft-versus-host (GVH) reaction which is a risk following bone marrow transplants.

Lesson outcomes

You should understand what stem cells are, and the uses of stem cell therapy.

WorkSpace

Dale
Personal health trainer

Scientists think that everyone should aim to do about 5 hours a week of moderate to strenuous exercise. It needs to be aerobic exercise – something that increases your pulse and breathing rates. This will use fat and help with weight control.

There should also be some anaerobic exercise, such as press ups, sit ups and weightlifting, as this helps strengthen and tone your muscles.

Sometimes I work with clients in the gym and help them develop an exercise programme that they can sustain and enjoy. If they enjoy it they are more likely

to exercise regularly. I also work with clients who do not like going to a gym. There are many things that people can do, such as swimming, walking, cycling, skipping, playing team sports, dancing and Zumba®, to exercise often and regularly.

Before anyone starts an exercise programme, I give them a health check. I find their weight, height, resting blood pressure, resting pulse rate (which shows their resting heart rate) and resting breathing rate. I find out if they have any health problems or mobility issues. I also need to take into consideration their age and lifestyle. Together we design an exercise programme that they can stick to and enjoy. If they are not fit, we start with a moderate level of exercise, and after a few weeks we measure resting blood pressure and pulse rate and can see if there has been some improvement. We can then increase the duration and intensity of the exercise.

Measuring the difference between resting pulse rate and pulse rate after exercise is also a useful way of determining fitness.

For anyone who cannot walk, they may be able to swim, use a rowing machine or take part in paraplegic sports, which develops their upper body strength and general cardiovascular fitness.

Think about it

1 When Dale measures his clients' resting pulse rates and blood pressure he takes three readings and finds the average. Why do you think he does this?

2 An indication that your cardiovascular fitness has improved is a drop in your resting heart rate (resting pulse rate).
   (a) Why do your think your resting heart rate drops if you have better cardiovascular fitness?
   (b) Explain why your pulse rate indicates your heart rate.

Introduction

When you carry out investigations in your school or college laboratory you are using scientific skills. You make predictions and you plan ways of testing your predictions. These are the same investigative skills used by those who work in the scientific industry.

In this unit you will learn how applied scientists and technicians use their knowledge to make predictions and develop methods that will give them reproducible and valid data. You will look at data that can be obtained in the workplace and you will develop skills to help you analyse the data to draw conclusions and test a hypothesis.

Someone working in a science-based job must always consider how certain they are about their conclusions; you will learn about the questions they ask themselves to help them evaluate their investigations and find ways of improving their evidence. For example, a radiographer looking at the results of a scan must decide whether the information is clear enough to make a diagnosis. If not, they may decide to take the scan from a different angle.

The investigations that you will read about in this chapter all link back to ideas you have met in Units 5, 6 and 7, helping you to see even more clearly how these ideas are used by working scientists and technicians.

Assessment: You will be assessed externally using a paper-based test.

Learning aims

After completing this unit you should:

a understand how to produce a good plan for an investigation

b be able to process, present and analyse data, and draw evidence-based conclusions

c be able to evaluate evidence and investigative methods.

"It was really interesting to see the connection between the experiments we do at school and what happens in the real world of working scientists and technicians.

I hadn't realised before why it is so important to plan and evaluate investigations so carefully, and that this is a really important part of a scientist's or a technician's work.

Marcus, *16 years old*

Scientific Skills

8

8.1 Scientific investigations

Get started

Have you carried out any investigations in this course? Discuss in pairs and try to think of three investigations. What were you trying to find out in each one?

Key terms

Hypothesis – A statement that can be tested by scientific investigation. The plural of hypothesis is hypotheses.

Working scientists and technicians need to carry out investigations for many reasons.

1 They may need to test the properties of new substances or products – for example an electrical engineer investigates how the resistance of a new type of thermistor depends on the temperature.

2 They may need to find the most efficient way of making a new substance – for example a chemical engineer investigates the most efficient way of manufacturing a nanochemical used in a cosmetic.

3 They may need to find out how effective a medical treatment is – for example a medical researcher investigates whether a new antibiotic is more effective than the current best treatment.

4 They may need to test samples of a substance to check what it contains – for example a laboratory technician investigates water samples in a river for signs of pollution.

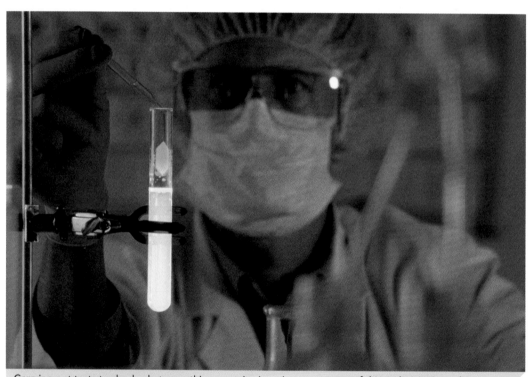

Carrying out tests to check what something contains is an important part of this technician's work.

What needs to be found out?

Any investigation starts with a question. The question must make it clear what the scientist or technician is trying to find out.

At this stage, a scientist will do some research. They will find information that is relevant to the question and use it to decide on a **hypothesis** and a plan.

Other scientists may have already carried out similar investigations. These may be used to help the scientist write their plan.

Activity A

Look at the four investigations described at the start of this lesson. In each case, write down the question that the scientist or technician is trying to answer.

How am I going to find it out?

Planning an investigation can be a long and detailed process. In the next few lessons you will look at each aspect of this process in more detail.

In their plan, scientists need to decide on:

- a suitable hypothesis that will be tested in the investigation – see lesson 8.2
- the equipment and materials that are needed – see lesson 8.3
- the factors that must always be the same for measurements to be compared – see lesson 8.4
- what risks might be present in the practical work and how they can be reduced – see lesson 8.5
- the measurements that will be taken – see lesson 8.6.

Case study

My name is Natalie and I work as a research chemist for a company developing nanotubes for use as fibres in protective clothing and sports gear. My job is to develop a new method to produce nanotubes. We are trying to grow them on a surface dotted with tiny crystals of metal catalyst. When this process is used in the factory it will be on a large scale, capable of generating several tonnes of nanotubes per week. But at the moment we are running a series of pilot studies on a much smaller scale. That way, we can change all sorts of factors, like the size of the catalyst crystals and their temperature. We can then see how those changes affect the rate and yield of the process, as well as the properties of the nanotubes. When we've finished these studies, the chemical engineers will scale the process up. We will need to work quickly to get ahead of our competitors.

1 Give some reasons why companies will run pilot studies before they commit to using a new chemical process like this one.

2 Many pilot studies do not lead to a new process being introduced. Suggest reasons why.

3 Apart from rate and yield of the process, what other information from the pilot study might Natalie's company be interested in?

Just checking

1 Look at the following four descriptions. Which one is the best description of a hypothesis?
  (a) The title of the investigation you are doing.
  (b) The question which you are hoping to answer.
  (c) A prediction which you can test using experiments.
  (d) The plan which you write down for your investigation.

Link

You learned about nanochemicals in lessons 5.15 and 5.16.

Did you know?

Threads created from super-tough carbon nanotubes can be used to make bullet-proof vests as light as t-shirts. The nanotubes make the vests more effective and less bulky than this older-style vest.

Lesson outcome

You should know the meaning of the term hypothesis.

8.2 Making a hypothesis

Get started

A baker places an advertising board in his shop window which reads: 'Our bread is the best in town'. Could this statement be called a hypothesis? Discuss in pairs.

Key terms

Qualitative – Describes things that can be observed, without using numbers.

Quantitative – Describes something using numbers.

Does eating too much red meat affect your chances of developing cancer?

The important thing to remember about a hypothesis is that it must be possible to test it using scientific investigation. It is a statement that states clearly what the scientist expects to happen in the investigation, based on relevant scientific theory.

Epidemiologists study the factors that cause diseases. A team of epidemiologists is carrying out research into the effect of eating red meat on bowel cancer.

They could make two kinds of hypothesis: **qualitative** or **quantitative**.

'Eating more red meat will increase the chance of developing bowel cancer.'	This is a qualitative hypothesis – it describes what they expect to find without using numbers.
'Eating more than five portions of red meat a week will increase the risk of developing bowel cancer by more than 10%.'	This is a quantitative hypothesis – it describes what they expect to find by using numbers.

Engineers and physicists will often use equations to describe how different substances and systems behave. This helps them to make *quantitative* hypotheses.

Activity A

Why do you think it is much harder for biologists or health workers to make quantitative hypotheses than it is for physicists and engineers?

Imagine a new source of crude oil has been found. You are a laboratory technician who is analysing it by distilling off different fractions of the oil. What is your hypothesis about the size of the molecules which will be distilled off at low temperatures?

What kind of molecules will be distilled off from this sample?

Assessment tip

You may be asked to write a hypothesis for an investigation. You must be able to justify the hypothesis using scientific ideas from Units 5–7. There are opportunities in this lesson to help you practise this skill.

- You know that smaller molecules have weaker forces between molecules.
- This means that they have lower boiling points.
- So you can state that the molecules that are distilled off at low temperatures will be the smallest molecules.

This is a *qualitative* hypothesis because no numbers are involved.

What if a hypothesis is wrong?

Is it always bad news if your hypothesis turns out to be wrong? It depends on your point of view. If you are a scientist who has developed a new nanowire that you are hoping will be much stronger than other nanowires, then it may be disappointing if your hypothesis is proved to be incorrect.

But you might be working in quality control. If the new nanowire is about to go into production and you suddenly find it is not as strong as it should be, then there is obviously a problem in the production process. This will need immediate attention. So the negative result is really important.

Many of the most important scientific breakthroughs have been made when a confident hypothesis turned out to be incorrect. Scientists develop new and original theories to explain the results, and develop a hypothesis of their own to test. If their hypothesis is correct then these new theories will become accepted by other scientists.

Did you know?

Scientists once thought smoking did not cause any harm to human health. Then, in the 1950s, it was discovered that smokers had a much higher chance of developing lung cancer than non-smokers. As a result of this, further studies into the effects of smoking were carried out and we now know that smoking can greatly increase the risk of developing many diseases, such as heart disease, emphysema and several types of cancer.

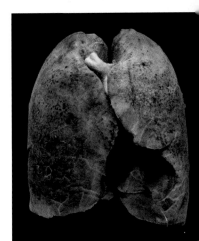

Scientific studies have shown that smoking may cause several serious diseases. The picture above shows the effect that smoking can have on human lungs.

Just checking

1 You are an electrician and intend to connect two lighting circuits in parallel. Each circuit has a resistance of $100\,\Omega$.
   (a) What is your hypothesis about the resistance of the two circuits when they are combined?
   (b) Is your hypothesis quantitative or qualitative?

2 Look at hypotheses (a) to (e) below. Decide whether each one is qualitative or quantitative.
   (a) The resistance of this thermistor will decrease when the temperature is increased.
   (b) The energy output per gram of bioethanol fuel is 50% lower than that of petrol.
   (c) The stopping distance will quadruple if the speed of a car is doubled.
   (d) People who are obese (with a BMI of 30–35) have a life expectancy 5 years shorter than those of average weight.
   (e) This water sample is polluted.

Lesson outcome

You should know how to produce quantitative and qualitative hypotheses.

Get started

If you were going to measure 100 cm³ of water in an experiment, would you use a beaker or a measuring cylinder? Why did you make this choice?

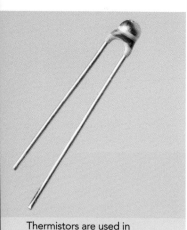

Thermistors are used in temperature-sensing systems.

Link

You can remind yourself about thermistors and how to measure their resistance in lesson 6.23.

Nadia is an electronics engineer. She is about to test a new type of thermistor for use in a temperature sensor. She needs to find out the characteristics of this thermistor. The designer of the sensor needs a thermistor that will be very sensitive to the temperature changes found in a building – which will be in the range 18–30 °C.

As the temperature changes, the resistance of the thermistor will change – Nadia's job is to find out exactly how it will change and write a report on her work.

Nadia's report should do two things:

1 give the results of her investigation

2 explain how she carried out the investigation so that other engineers can check that they can get the same results.

Measuring resistance

Nadia does some research to find out how to test thermistors. She learns that she can find out the resistance by measuring the current through the thermistor when a voltage is applied.

She will need to measure the voltage and the current. She will use a voltmeter and am meter to do this.

Activity A

1 Draw a diagram to show a circuit that Nadia could use to measure the resistance of a thermistor at room temperature.

2 List all the equipment that will be necessary to set this circuit up.

Changing the temperature

Nadia finds out that the best way of changing the temperature of the thermistor is to place it in a beaker of water. If she starts with a beaker full of hot water she can let it gradually cool and take readings of voltage and current at different temperatures.

She will use a thermometer, and she chooses to use one that measures in the range 0–50 °C as this includes the range of temperatures she is interested in. Because it has a small range it has markings every 0.1 °C, so it can be used to measure temperature to a greater **precision**.

This thermometer can show temperature to the nearest 0.1 °C.

Safety and hazards

Nadia's experiment requires the use of water and a power source. If you carry out the experiment yourself, make sure that: the beaker of water is stable when you place the thermistor inside it; you do not touch the power source with wet hands; and you keep the power source as far away from the beaker as possible.

Explaining the choice of equipment

In her report, Nadia could explain why she chose each piece of equipment.

Equipment chosen	Further information about the equipment chosen	Reasons for choice
Voltmeter	Multimeter Reads from 0.4 V to 100 V Displays three digits (for example 5.25 V)	It will measure the voltage across the thermistor. It will give a precise measurement.
Ammeter	Multimeter Reads from 0.04 A to 10 A Displays three digits (for example 0.125 A)	It will measure the current through the thermistor. It can measure current over a wide range. It will give a precise measurement.
Thermometer	Reads from 0 °C to 50 °C, in 0.1 °C intervals	It will measure the temperature of the water. It has a suitable range. It will give a precise measurement.

Take it further

When investigating circuits, scientists and technicians use digital multimeters to take measurements. These can measure voltage, current or resistance. They can measure very wide ranges of values to a high level of precision.

A digital multimeter.

Did you know?

You probably think the balances in your school or college laboratory are precise because they measure the mass of an object to the nearest 0.1g or 0.01g. But the most precise piece of equipment used to measure mass is the Penning Trap Mass Spectrometer – this can allow scientists to make measurements of the mass of individual atoms to the nearest 0.000000000000000000000000000001 g. It's great for investigating atoms but it wouldn't be much use if you were trying to weigh a lump of metal!

Activity B

1 List all the equipment Nadia will need in order to *change* the temperature of the thermistor.
2 Explain the reasons for using each piece of equipment.

Link

Look back to lesson 5.3 to remind yourself how to calculate how much energy is given out by a fuel.

Just checking

1 You are an applied chemist working for an oil refining company investigating the energy given out by different liquid fuels. Write a list of the equipment you would need to carry out the investigation.
2 Give reasons for your equipment choices in question 1.

Lesson outcomes

You should be able to identify relevant equipment for an investigation and give reasons for your choices.

8.4 Variables

Get started

Think about some experiments that you have done in this course. What kinds of things have you measured during those experiments?

Key terms

Dependent variable – A variable that is measured for each and every change in the independent variable.

Independent variable – A variable that is purposely changed during an investigation.

Activity A

Use scientific ideas from lesson 7.2 to justify Darryl's hypothesis.

Did you know?

Many studies of the factors affecting health take place over a very long time. One study in the USA looked at the health of 100 000 nurses for 35 years. It has provided valuable evidence about the effect on health of diet, smoking and alcohol use.

Assessment tip

You need to be able to identify the dependent and independent variables in an investigation.

Darryl works in a team of health workers. The team is carrying out some research into the link between the amount of exercise you do and the risk of having a heart attack. Darryl's hypothesis is that the less exercise you take, the more likely you are to have a heart attack.

The team monitors 1000 men aged between 55 and 65 for a period of 10 years. They ask each person about the amount of exercise they take each week.

Finding out about someone's lifestyle is an important part of health research.

Dependent and independent variables

Things that are measured or changed in an investigation are called variables. There are different types of variable.

The **independent variable** is one that the experimenter deliberately changes (or allows to change) during the investigation.

The **dependent variable** is one that may change as a result of changes to the independent variable; changes in the dependent variable are what is measured during the investigation.

In Darryl's research, the *independent variable* is the amount of exercise taken. Darryl has chosen men who take different amounts of exercise so he has allowed this variable to change.

Darryl plans to divide the men taking part in the study into three groups: those taking no exercise, those taking some exercise and those taking a lot of exercise.

The *dependent* variable in Darryl's research is the chance of having a heart attack. Darryl will try and measure this by looking at the number of men in each group who have a heart attack during the 10 years of the study.

The table shows Darryl's results.

Group	Amount of exercise per week	Number of people in the group	Number of heart attacks in a 10-year period
1	None	231	12
2	Less than 60 minutes	430	14
3	More than 60 minutes	339	4

Activity B

1 Look at Darryl's results. Which group of men had the most heart attacks?
2 Calculate the percentage of each group who had a heart attack during this 10-year period. Which group had the highest percentage of heart attacks?
3 Which of these two ways of comparing the results do you think is the fairest way? Explain your answer.

Will exercising like this reduce his chance of having a heart attack?

Fair test and control variables

A **control variable** is one that must be kept constant during the investigation if it is to be a fair test.

An investigation will be a **fair test** of the hypothesis if the independent variable is the only thing allowed to change.

Darryl knows that smoking is another factor that can affect your chance of having a heart attack. So he makes sure that each of the groups in his study has an equal number of smokers and non-smokers.

Any variable that can affect the chances of having a heart attack will be a control variable in this investigation. Darryl will also need to take account of several other control variables when he chooses which men to follow in this study.

Activity C

Think about other factors that can affect the chances of having a heart attack.

1 Write a list of these other control variables.
2 Choose one of these variables and suggest what Darryl could do to make sure that his study is still a fair test.

Link

Look back to lesson 6.22 to remind yourself of the properties of light-dependent resistors.

Just checking

1 Here are four questions which could be investigated. For each question, identify the dependent and independent variables.
   (a) What happens to the resistance of a light-dependent resistor when you shine a light on it?
   (b) Do alkanes with a high boiling point release more energy when they are burnt?
   (c) Are rivers flowing through towns more polluted than rivers flowing through countryside?
   (d) How does the speed of a car affect its stopping distance?

Lesson outcomes

You should be able to identify dependent and independent variables and describe how they are controlled.

8.5 Managing risks

Get started

What hazards do you need to be aware of when you do practical work? What precautions do you take to protect yourself against these hazards?

Key terms

Risk – The harm that could be caused and the chances of it happening.

Safety and hazards

Different workplaces will use different sources of information about hazards, risks and control measures.

In your school or college laboratory, the CLEAPSS database is the usual source of information for a risk assessment.

Vijay is a health and safety officer for a company manufacturing paint strippers. His job is to make sure that the technicians in the laboratories work in a safe and healthy environment, and that government legislation on health and safety is followed.

A group of technicians is investigating how effective a new paint-stripping product is. The paint stripper contains a number of chemicals including dichloromethane (DCM).

The risks of using paint stripper must be assessed, even when it is used in a scientific laboratory.

Risk assessments

Vijay will carry out a **risk** assessment on the procedure. To do this he will:
- look at the laboratory and the method being used in the investigation
- identify any hazardous substances or any activities (like handling hot liquids) that could cause harm
- consider the risks involved in using these substances
- evaluate the best way to manage and control the risks.

Activity A

Look at this extract from Vijay's risk assessment. Match each control measure (**A–C**) with the risk (**1–3**) that needs to be reduced.

Hazard	Risk	Control measures taken to reduce the risk
Dichloromethane is harmful.	1 Can cause damage to nervous system if inhaled 2 Can cause burns if in contact with skin 3 Long-term exposure may cause cancer	A Wear skin and eye protection B Limit exposure to DCM to 2 hours in any 24-hour period, and to 1 month in every 12-month period C Work in a fume cupboard or well-ventilated area

Other sources of risk

In many scientific investigations, the use of chemicals is the main source of risk, but there are other risks that may need to be considered, such as those shown in the table below.

Hazard	Risk
Handling hot objects	Burns
Ionising radiation (from radioactive sources)	Various forms of cancer
UV (ultraviolet) radiation	Eye damage, skin cancer
Handling microorganisms and wildlife	Diseases

Medical researchers may be exposed to dangerous microorganisms in their investigations.

Activity B

Suggest some control measures you would take to minimise the risks of:
1 exposure to UV radiation
2 handling microorganisms.

Information about hazards

If a chemical is a potential **hazard**, it must be labelled with the correct hazard symbol. Anyone using a substance with a hazard label should use the CLEAPSS Student Safety Sheets to find out more details about the risks.

Activity C

Use a set of CLEAPSS Student Safety Sheets to find what control measures you should take if you needed to handle:
1 a flammable substance
2 a corrosive substance.

 Explosive

 Flammable

 Corrosive

Hazard symbols warn you of the potential risks of handling chemical substances.

 Link

Look back to lessons 5.3 and 8.3 to remind yourself how to calculate how much energy is given out by a fuel.

 Lesson outcomes

You should be able to identify risks in an investigation, and carry out a risk assessment to explain how they will be managed.

Just checking

1 Use a set of CLEAPSS Student Safety Sheets to find out what labels should be attached to bottles containing:
   (a) concentrated sulfuric acid
   (b) ethanol
   (c) lead(II) nitrate.
2 You are a chemical technician working for an oil refining company investigating the energy given out by different liquid fuels. List the risks involved in this experiment and the control measures that you would use to reduce the risks.

8.6 Decisions about measurements

Key terms

Accurate – An accurate measurement is close to the true value.

Anomalous (anomaly) – A result is anomalous if it falls outside the normal (or expected) range of measurements.

Link

You will learn more about identifying proportional patterns in data in lesson 8.15.

Look back to lesson 8.2 to remind yourself about quantitative hypotheses.

Colette is a chemical engineer investigating the best conditions for manufacturing chloromethane, a useful solvent. She is studying the reaction of hydrochloric acid with methanol to make chloromethane and water. She now wants to find out how the rate of the reaction is affected by the concentration of the hydrochloric acid (HCl).

She has already decided how to measure the rate of reaction, but now needs to make some more decisions about measurements.

Colette is investigating an organic reaction.

How many measurements?

Colette has a hypothesis: the rate of the reaction will increase as the concentration of HCl increases. In fact, she expects that when she doubles the concentration of HCl, the rate of reaction will double. The two variables, rate and concentration, are directly proportional to each other.

- If she only uses two concentrations she will not have enough information to be sure about a quantitative hypothesis.

- Colette knows that the best way of testing a quantitative hypothesis like this is to plot a line graph.

- Colette wants to include at least five points on her graph. This will mean she will be able to draw a best-fit line and ignore any **anomalies**.

What range of concentrations?

Colette now needs to decide exactly which concentrations to investigate.

- Colette knows that for a line graph to show a pattern clearly, the points on the graph should cover a wide range.

- The points need to be fairly evenly spaced.

- She also wants to choose concentrations that might be used in the actual manufacture of chloromethane.

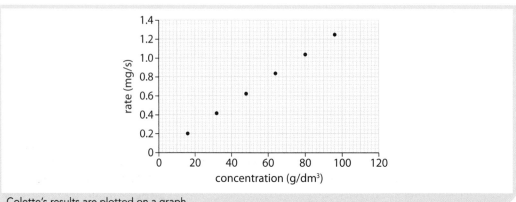
Colette's results are plotted on a graph.

1 How many different concentrations did Colette investigate?

2 Write down the actual concentrations of hydrochloric acid that she used.

Henry and Sinead, two trainees in Colette's team, make different decisions about the measurements to make.

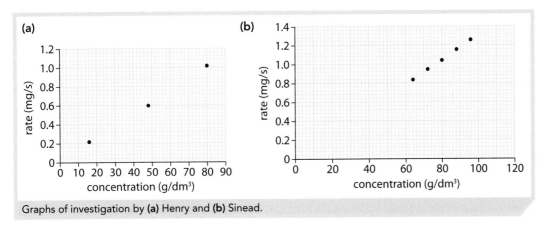

Graphs of investigation by **(a)** Henry and **(b)** Sinead.

Look at the graphs of Henry and Sinead's investigations. Write down the answers to these questions:

1 What is different about the decisions that they took, compared with Colette's investigation?

2 How would you improve their method?

Repeating measurements

In most investigations, a scientist or technician will try and repeat measurements, so that they are done at least three times. There are several reasons why they do this.

- If the results vary slightly it will allow them to calculate an average. Averages using several results are more **accurate** than a single value.

- Anomalous results can be identified. These can then be excluded from the calculation of the average.

Some measurements will need to be constantly repeated so that any changes can be quickly spotted. An analyst doing water sampling for the Environment Agency needs to test water regularly to monitor oxygen concentration. If the oxygen concentration falls to low levels, it may be a sign that there are serious environmental problems in rivers and lakes.

In these cases, scientists or technicians will need to decide how often the measurements should be made.

Link

You will learn more about anomalous results in lesson 8.9.

Just checking

1 Simon is a fitness instructor. He is making an assessment of a client's fitness and wants to monitor her pulse rate during and after a 10-minute exercise session. Suggest when Simon should make measurements of his client's pulse.

2 Gemma is a road safety officer. She wants to investigate how stopping distance is affected by the speed of a car to help her make recommendations about the speed limit on a road outside a school. Suggest a suitable range and number of speeds for her to investigate.

Lesson outcomes

You should be able to give a suitable range and number of measurements and explain your decisions.

8.7 Writing a plan

Get started

When you do an investigation, you usually write a plan before you start. Why is it important for you to do this?

Link

You learned about PKU in lesson 7.9.

Sunita works as a hospital laboratory technician. She is responsible for carrying out Guthrie tests. Guthrie tests are done on newborn babies to check for a rare but serious disease called phenylketonuria (PKU).

A midwife takes a drop of blood from the baby's heel and spots it onto filter paper. Sunita then tests a small piece of the filter paper to see if it promotes the growth of bacteria. If it does, then the blood contains high levels of phenylalanine which means that the baby may have PKU.

If the results are to be repeatable the test must be done in the same way, every time.

Laboratory method

This is Sunita's plan for testing a batch of 10 samples.

1 Add 50 g of agar to 100 cm^3 of boiling water and cool to 50 °C.
2 Add a **suspension** of *Bacillus subtilis* to the agar solution.
3 Pour the mixture into plates and allow to set for 30 minutes.
4 Cool to 4 °C.
5 Remove a plate 1 hour before use and leave at room temperature.
6 Measure a 3-mm-diameter disc of the blood spot from each baby using a digital **calliper**.
7 Place the discs on the surface of the plate and incubate in an oven at 37 °C for 18 hours.
8 Measure the diameter of the bacterial growth with a digital calliper. If the diameter of the zone is larger than 8 mm, then the test is positive.

Safety and hazards

Any worker who is handling microorganisms must follow strict guidelines to make sure that the possible hazards are controlled. These include:

- sterilising all equipment before and after use, by heating to a high temperature in an autoclave;
- washing hands before and after handling any equipment;
- disinfecting all work surfaces before use;
- wearing appropriate protective equipment such as safety goggles, gloves and face mask.

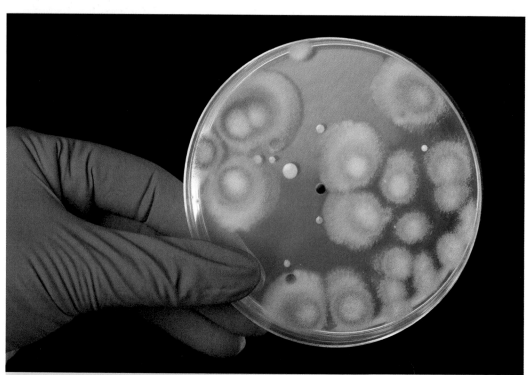

Bacteria growing on an agar plate indicate high levels of phenylalanine.

What makes a good plan?

A good plan must be able to be followed by other scientists. It must:

- explain the hypothesis that is being tested – *in this case the hypothesis is that if a baby has PKU, the blood sample will cause the growth of bacteria*
- give details of the equipment to be used, especially any measuring instruments
- explain which variable is being changed (the independent variable) – *in this case it is the source of the blood*
- suggest a suitable number and range of measurements to take – *Sunita took 10 measurements from different sources*
- list all the variables which must be controlled
- present the instructions for each step of the method in a logical order.

Why do scientists publish their results?

When scientists have obtained results from an investigation they will often publish their results and their method in a scientific journal.

If other scientists get the same results then it shows that the results are **reproducible**. This means that any **conclusion** is much more likely to be correct.

Activity A

1. Use ideas from lesson 7.9 to justify the hypothesis in Sunita's plan.
2. Write down any pieces of equipment that are included in Sunita's method. Are there any other pieces of equipment that Sunita will probably use, but which she has not mentioned?
3. What variables does Sunita control when she carries out these tests? Choose one of them and explain why it is important to control this.

Did you know?

In 1989, Stanley Pons and Martin Fleischmann told the world that they had found a solution to the world's energy problem. They claimed they had produced huge amounts of heat from water by a process called cold fusion. They published their results and details of their method, which involved electrolysis of heavy water on the surface of a palladium electrode. Other scientists tried the same experiment but none of them could get the same results that Pons and Fleischmann had claimed. There is still controversy within the scientific community over whether or not cold fusion works.

Just checking ✔

You are a chemist testing the hypothesis that when liquid fuels are burnt, the longer the hydrocarbon chain, the less energy per gram is produced.

Write a plan to investigate this hypothesis.

Hint: You may want to look back at the answers to the 'Just checking' questions in lessons 8.3 and 8.5.

Lesson outcome

You should be able to outline a logically ordered plan to test a given hypothesis.

Leo's job is to design rollercoasters like this one.

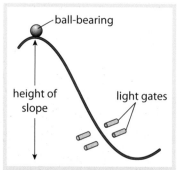

Leo uses this apparatus to test his rollercoaster design.

Leo is an engineer working for a company that designs theme park rides. He is working on a new design for a rollercoaster.

He is testing the speeds at the bottom of one of the slopes. He builds a model of the ride and drops a ball-bearing down the slope. Using light gates, Leo finds the speed of the ball-bearing at the bottom of the slope. He changes the height of the slope and investigates how this affects the speed of the ball-bearing. The length of the slope remains the same.

Activity A

Write a hypothesis for Leo's investigation.

Designing a table

Leo needs to write a report on his investigation. It is very important that the information in his report is easy to understand for anyone reading it. He decides to record his results in a table.

These are the features of a well-designed table:

- data arranged in columns
- each column has a clear heading to explain what it contains
- the heading for the column includes units, if necessary – some types of data do not have units
- the independent variable is in the first column.

Activity B

1 Look at Leo's table of results (right). Identify the independent variable and the dependent variable.
2 Has Leo chosen a suitable number and range of values of the independent variable to investigate? Explain your answer.

Height of slope (m)	Speed at bottom of slope (m/s)
0.40	2.52
0.80	3.56
1.20	4.34
1.60	5.01
2.00	5.59

Identifying patterns from tables

Leo could plot a graph of his results to help him interpret the data. But his well-designed table already helps him to see some of the patterns in the data.

The data for the smallest height appear at the top, and the height of the slope increases as you go down the first column. This is another important feature of a well-designed table:

- data should be arranged in order of the independent variable, this is usually in ascending (increasing) order.

Other types of data

Edith is a health worker in a large village in Uganda. She wants to find out how many villagers belong to each of the four main blood groups. She conducts a survey by testing 700 villagers. This table shows Edith's results.

Blood group	Number of people in the sample with this blood group	Percentage of the sample with this blood group
A	175	25
B	70	
AB	420	
O	35	

- The independent variable this time is the blood group.
- With surveys like this, it is very helpful to present the results as percentages. This means that they can be compared with the results of other surveys, even if the numbers of people in the different surveys are not the same.

Activity D

1 Calculate the percentages of people with blood groups B, AB and O in Edith's survey.

2 Suggest one reason why it might be important for health workers to know the percentages of people with each blood group.

Just checking

Salim works for a chemical processing company and has been asked to investigate the use of sodium hydroxide to neutralise acids. He investigates the temperature change when different volumes of acid are added to 25 cm³ of sodium hydroxide. He makes a table of his results.

List five things that are wrong with Salim's table.

°C	Acid added
4	5
7	30
12	10
14	20
21	15

Activity C

Look at the table in Activity B. Is it easy to see how the speed of the ball-bearing changes as the height of the slope changes? Write down the pattern that you observe from the data in the table.

This sample of blood will be analysed to find the person's blood group.

Link

You learned about the different blood groups in lesson 7.15.

Lesson outcome

You should be able to display data in a clear and logically organised table.

161

8.9 Handling data and anomalies

Get started

If you repeat an experiment several times, the measurements you make might be slightly different each time. Give some reasons why this might happen.

Paula is a technician working for an oil company. The company is developing some new lubricating oils for a customer who requires a product with a particular viscosity. Viscosity measures how easily a liquid flows.

Paula carries out an investigation in which she drops ball-bearings down a tube filled with oil. She uses light gates to measure the time taken for the ball to drop 0.5 m. The longer a ball takes to fall, the more viscous the oil must be. She repeats the test three times with each sample of oil.

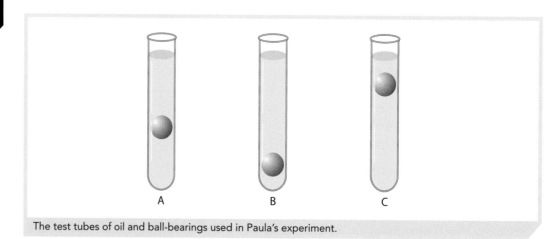

The test tubes of oil and ball-bearings used in Paula's experiment.

Link

You will learn how to recognise and handle anomalies in graphs in lesson 8.14.

The viscosity of lubricating oils can be measured by experimental investigation.

Anomalies in tables of data

The table shows Paula's results.

Oil	Time taken for ball to drop (seconds)		
	1	2	3
A	8.14	8.20	9.60
B	5.76	5.82	5.83
C	14.20	15.02	14.27

- Look at the results for oil B. All the times are within a narrow range – in this case within about 0.1 of a second. Results like this are called **concordant** results.

- Now look at the results for oil A. The first two results are close together, so they are concordant. The third result is different to these results by over a second. It falls outside the range of the rest of the results, so it is called an anomalous result – or sometimes just 'an anomaly'.

- Repeating an experiment helps you to see if there are anomalous results.

Activity A

The results for oil C also contain an anomalous result.
Which result is anomalous?

Calculating the mean

It is quite usual for measurements to vary slightly every time an experiment is done.

- To take account of this, scientists will repeat an experiment several times to get a number of measurements under exactly the same conditions.
- They can then calculate the **mean** (average) of the concordant measurements.
- Any anomalous results should not be included in the calculation of the mean.

Worked example

Calculate the mean time for oil B in Paula's results.

Step 1 Add up the times from all the tests: 5.76 + 5.82 + 5.83 = 17.41

Step 2 Divide the result by the number of measurements of time, which is 3:

$$\frac{17.41}{3} = 5.8033333$$

Step 3 The number is rounded down to 5.80, which is the same level of accuracy as the original measurements.

Activity B

1 Calculate the mean of the results for oils A and C.

2 Which oil is the most viscous? Explain your answer.

Just checking

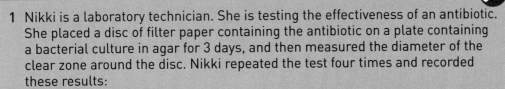

1 Nikki is a laboratory technician. She is testing the effectiveness of an antibiotic. She placed a disc of filter paper containing the antibiotic on a plate containing a bacterial culture in agar for 3 days, and then measured the diameter of the clear zone around the disc. Nikki repeated the test four times and recorded these results:

   4.4 cm, 5.2 cm, 0.7 cm, 4.9 cm

   (a) What does the clear area around the disc indicate?

   (b) Calculate an appropriate mean result for this antibiotic.

2 Sam is testing two different fertilisers, A and B, to see how they affect plant growth. He treats the plants with one of the fertilisers then measures how much the plants have grown after 10 days. The table shows his results.

Height after 10 days (cm)	6	8	10	12	14	16	18
Number of plants treated with A	2	4	5	3	3	1	2
Number of plants treated with B	0	1	2	4	6	5	2

   (a) Calculate the mean height for each group of plants.

   (b) Which fertiliser, A or B, produced the greatest mean plant height?

Take it further

There are actually three meanings of the word 'average': the mean; the median – the middle value when readings are arranged in ascending order; and the mode – the most common value.

Link

You will find out more about how to write the answers to calculations to a suitable level of accuracy, 'significant figures', in lesson 8.10.

Remember

Anomalous results should not be included in the calculation of a mean.

Lesson outcomes

You should be able to identify anomalous results in a table of data and calculate mean values of the measurements.

Get started

Very often in scientific experiments, you need to use the measurements you make to calculate other important quantities. What measurements would you need to make so that you could calculate:

(a) the speed of a moving object

(b) the resistance of a wire

(c) the heat energy absorbed by a beaker of water?

Key term

Significant figures – The digits in a number which give information about the precise value of that number.

Link

This lesson links with ideas about GPE and KE in lesson 6.5.

Look back at lesson 8.8 to remind yourself of Leo's first experiment.

Assessment tip

You will be given any equations that you need to use to answer a question. You should be familiar with the equations you met in Units 5, 6 and 7, as these are the equations you are most likely to need to use.

Remember

When you give the answer to a calculation, you must give the correct units as well.

Leo is carrying out further tests on his model of a rollercoaster ride. He uses the speed of the ball-bearing to calculate the kinetic energy (KE) at the bottom of the ride. He then compares this to the gravitational potential energy (GPE) which the ball loses as it falls down the steep slope.

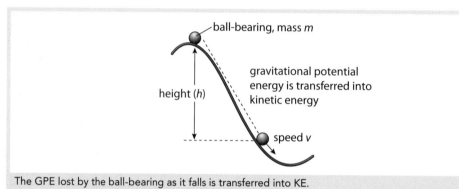

The GPE lost by the ball-bearing as it falls is transferred into KE.

Using equations

Leo measures the speed of the ball-bearing, so he can calculate its KE using this equation:

$$\text{KE (J)} = \tfrac{1}{2} \times \text{mass (kg)} \times \text{speed}^2 \text{ (m/s)}$$

He also measures the change in height of the ball-bearing, and he can calculate the GPE lost using this equation:

$$\text{GPE (J)} = \text{mass (kg)} \times \text{acceleration due to gravity (m/s}^2\text{)} \times \text{change in height (m)}$$

Finally, he measures the mass of the ball-bearing. It is 0.053 kg.

Leo designs a table, with space for him to include the results of his calculations. He uses a value of 10 m/s² for the acceleration due to gravity.

Height of slope (m)	Speed at bottom of slope (m/s)	GPE lost by ball-bearing (J)	KE of ball-bearing at bottom of slope (J)
0.40	2.52	0.212	0.168
0.80	3.56		
1.20	4.34		
1.60	5.01		
2.00	5.59		

Activity A

Make a copy of Leo's table. Calculate the missing values for the GPE and KE.

Significant figures

When you use data from experiments in calculations, you will sometimes get results on your calculator that include many digits.

A car travels 29 m in 7.0 s. You calculate its speed and your calculator gives you this answer:

You measured the distance only to the nearest metre and the time to the nearest 0.1 of a second, so there will be a limit to the accuracy of the answer you calculate.

The key to writing your answer is thinking about **significant figures**. You should write your answer using the same number of significant figures as the data. This may mean you have to round a number up or down.

In this example:

- 29 m has two significant figures
- 7.0 s has two significant figures; you measured the time to the nearest 0.1 s, so the zero does actually mean something.

Your calculation for the speed should be rounded down to two significant figures. You write it as: 4.1 m/s.

Other equations

Here are some of the other equations you might need to use in your test, with a lesson number for you to look up more details.

Equation	Lesson
Heat energy (J) = mass of water (g) × specific heat capacity (J/K/g) × temperature change (°C)	5.3
Distance (m) = speed (m/s) × time (s)	6.2
Acceleration (m/s²) = $\dfrac{\text{change in velocity (m/s)}}{\text{time taken (s)}}$	6.3
Force (N) = mass (kg) × acceleration (m/s²)	6.11
Work done (J) = force (N) × distance (m)	6.12
Resistance (Ω) = $\dfrac{\text{voltage (V)}}{\text{current (A)}}$	6.21

Activity B

1 How many significant figures are shown on the calculator display on the left?
2 Look at this list of numbers: **(a)** 2.78; **(b)** 0.45; **(c)** 0.000 478; **(d)** 4 560 000. How many significant figures are shown in each case?
3 John calculates the speed of a car as 2.181 936 m/s. Rewrite this number to two significant figures.

Link

Refer back to lesson 5.3 to remind yourself how to calculate specific heat capacities.

Lesson outcome

You should be able to carry out calculations from data using equations, giving your answers to the correct level of accuracy and to an appropriate number of significant figures.

Just checking

James, a food technologist, burns a cashew nut and uses it to heat up 100 g of water. The temperature increases by 9.25 °C. Calculate the heat energy released, giving your answer to an appropriate number of significant figures. The specific heat capacity of water is 4.2 J/K/g.

8.11 Bar charts and pie charts

Get started

Bar charts and line graphs are two different ways of displaying data. What are the differences between the two kinds of graphs? Can you think of an experiment you have done when you displayed your results as a line graph? And one when you displayed your results as a bar chart?

Link

You learned about the different blood groups in lesson 7.15.

Remember

It is very important to make sure that you put labels on both axes of a graph. The labels must say exactly what data are being displayed. It is usually a good idea to use the labels at the top of the data table as the labels for the axes.

Often, the data in the table have units. These units must also be included in your labels.

Discussion point

Compare the pie chart and bar chart showing Edith's data. Which one do you prefer as a way of displaying the data? Give reasons for your choice.

You read about Edith, the health worker in Uganda, in lesson 8.8. Edith investigated the number of people in a village with different blood types. She is now writing a report on her investigation for the regional health council.

She wants to find a way of showing the patterns in her data clearly. Scientists and technicians usually plot a graph to do this. They plot different types of graph depending on the type of data they want to display.

Bar charts

The independent variable in Edith's investigation was blood group. There are four blood groups: A, B, AB and O. Edith decides that the best way to display this kind of data is to use a bar chart.

A bar chart uses the height of different columns to show the values of the dependent variable. The dependent variable is always plotted on the vertical axis (*y*-axis) and the independent variable is plotted on the horizontal axis (*x*-axis).

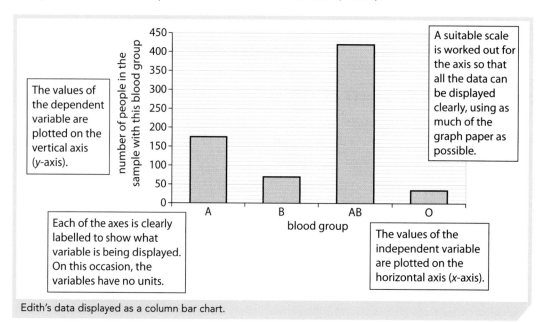

The values of the dependent variable are plotted on the vertical axis (*y*-axis).

A suitable scale is worked out for the axis so that all the data can be displayed clearly, using as much of the graph paper as possible.

Each of the axes is clearly labelled to show what variable is being displayed. On this occasion, the variables have no units.

The values of the independent variable are plotted on the horizontal axis (*x*-axis).

Edith's data displayed as a column bar chart.

Pie charts

Another way that Edith could display her data is to draw a pie chart. The value of the dependent variable is shown by the size of the sector.

Pie charts can only really be used in investigations such as surveys, where the total number of people in the survey is known. They can be useful when comparing the results of different surveys, even if the numbers of people taking part in the surveys are very different.

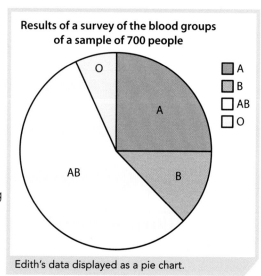

Results of a survey of the blood groups of a sample of 700 people

Edith's data displayed as a pie chart.

Worked example

To draw a pie chart you need to calculate the angles for the different sectors of the circle. The total of all the angles from each sector must add up to 360°. Calculate the angle for the A blood group sector in Edith's pie chart.

Step 1 There are 700 people in the survey, and 175 people have blood group A.

Step 2 So the angle for this sector will be: $\frac{175}{700} \times 360° = 90°$

Activity A

Look again at the data for Edith's investigation in lesson 8.8.

1 *Calculate* the angles for the sectors in the pie chart representing blood groups B, AB and O.

2 Now *measure* the angles in the pie chart on page 24. How closely do your measurements match up with your calculations?

Just checking

1 Table 1 gives the density of some organic liquids.
Display the data as a bar chart, including suitable labels for the axes.

Table 1	
Liquid	Density at 25 °C (g/cm³)
Ethanol	0.79
Glycerol	1.26
Ethanoic acid	1.05

2 Table 2 shows the results of a survey investigating the amount of exercise done by a sample of 60 school children.

Display the data as a pie chart, including suitable labels and title.

Table 2	
Time spent exercising per week (minutes)	Number of children
0–60	9
60–120	33
120–180	12
More than 180	6

Lesson outcome

You should be able to draw bar charts and pie charts from data provided for you.

8.12 Using line graphs

Get started

When you plot line graphs it is important to know the difference between a *dependent* and an *independent* variable. Discuss the difference between these two types of variable.

In lesson 8.6, Colette investigated how the rate of the reaction depended on the concentration of hydrochloric acid she used. Now, she wants to find out how the temperature affects the rate of reaction. She carries out a pilot study to investigate this. The results from Colette's investigation are shown in the table.

Temperature (°C)	Rate of reaction (g/min)
12	1.2
21	2.3
30	4.2
44	8.6
50	13.5

Activity A

Which is the dependent variable and which is the independent variable in Colette's investigation?

Link

Look back to lesson 8.11 to remind yourself which variable to put on which axis, and how to label your graph axes.

In Colette's investigation, both the dependent variable and the independent variable are in the form of numbers. The best way to display these results is to do a scatter graph. The data are plotted as a series of points 'scattered' over the graph. Colette decides on suitable scales for the *x*-axis and *y*-axis and then plots the points accurately.

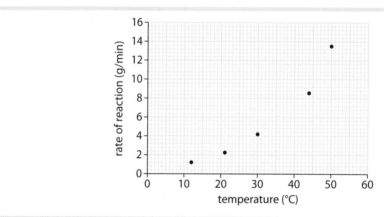

When both variables are in the form of numbers, the points can be plotted as a scatter graph. The pattern in the data can be shown by drawing a best-fit line.

Colette can now show the pattern in the data more clearly by drawing an appropriate line through the points. A completed scatter graph with a line added to show the pattern is called a **line graph**.

Assessment tip

It is really important to check that you have plotted all the points accurately. If some of the points look as though they do not fit the pattern, check that you have plotted them correctly. If they still do not fit the pattern then they are probably anomalies.

Remember

You need to make sure that the scale you use is regular. That is, the intervals between the numbers are the same all the way along each axis.

Best-fit lines

In most cases, scientists draw a **best-fit line** to show the pattern in the data.

- A best-fit line is a line drawn to pass as close to as many of the data points as possible.
- It can be a straight line or a simple curve.
- Any points that are clearly anomalies should not be included on the best-fit line.

Colette looks at the points on her graph. She tries to draw some best-fit straight lines, but only two or three of the points seem to lie close to a straight line. She concludes that a straight line is not the best way to show the pattern.

She then tries to imagine a best-fit curve. It looks to her as though all of the points will lie close to this curve. So she decides to draw a best-fit curve.

Other types of data

Andy is a researcher working for a health authority studying the numbers of lung cancer cases over a period of several years. He plots a graph to show his results. By joining the points with straight lines, he shows the pattern in the data.

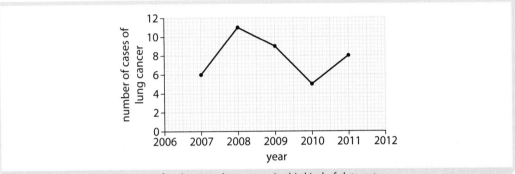

Best-fit lines are not appropriate for showing the pattern in this kind of data set.

In his study Andy measures the number of cases in a year. There is only one value of the dependent variable each year, so trying to draw a smooth best-fit line to show the pattern of what might happen between each point makes no sense.

Just checking

Gemma is an optometrist trying to find a suitable material to replace glass for a customer's lenses. She investigates how the angle of refraction depends on the angle of incidence for one possible material. Here are her results.

Angle of incidence (degrees)	Angle of refraction (degrees)
15	19
30	19
45	28
60	39
75	46

1 What is the independent variable in Gemma's investigation?

2 Plot the results on a suitable graph and draw an appropriate best-fit line to show the pattern in the data.

Lesson outcome

You should be able to draw line graphs with lines of best-fit.

8.13 Calculations from graphs

Get started

Explain how you can calculate the slope (gradient) of a line. Being able to do this is a very important skill when you are using line graphs.

Activity A

1 Use Nadia's graph to predict:

   (a) the current, when the temperature is 48 °C

   (b) the temperature at which the current would be 0.20 A.

2 Do you think that Nadia could accurately predict the current when the temperature is 4 °C?

You read about Nadia, an electronics engineer, in lesson 8.3. She is carrying out investigations into the characteristics of a new thermistor for use in a temperature sensor to monitor the temperature in a building.

In one investigation, Nadia applied a voltage of 6.0 V across the thermistor and measured the current at five different temperatures. Here is a graph of her results, after she has drawn a best-fit curved line.

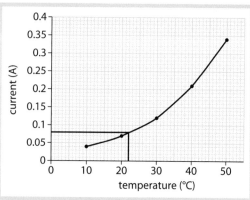

Nadia uses this graph to predict the value of the current at 22 °C.

Nadia wants to find the current at 22 °C. She does not have any data for this temperature, but she can use construction lines on the graph to make a prediction.

- She draws a line vertically up from the x-axis at 22 °C.
- When this meets the best-fit line she draws a line horizontally across to the y-axis.
- She reads the value of the current from the scale: it is 0.08 A.

Using data from graphs in calculations

Nadia needs to calculate the resistance of her thermistor at 22 °C, using the equation:

$$\text{resistance } (\Omega) = \frac{\text{voltage (V)}}{\text{current (A)}}$$

The voltage was 6.0 V in her experiment. She uses the value of the current at 22 °C:

$$\text{resistance} = \frac{6.0}{0.08} = 75\,\Omega$$

Using the slope of a graph

Umar is a car designer. He is investigating whether the cruise control on a new model of car will keep the speed of the car constant.

He sets the cruise control to 40 mph (miles per hour) and uses GPS to measure the distance in metres travelled by the car over a period of 100 seconds.

The graph shows Umar's results.

- Umar draws a best-fit straight line.

- He knows that if a graph of distance against time is a straight line, then it means that the speed is constant.

- He can calculate a value for the speed by finding the gradient (slope) of the line.

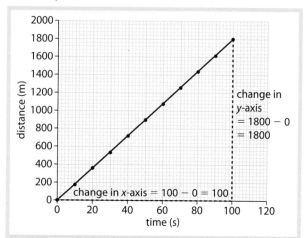

Umar can calculate the gradient of this straight line graph.

Did you know?

The GPS systems that are often fitted to cars are accurate to within about 20 m. More expensive systems can be accurate to within a few centimetres, although if they are being used to measure speed accurately (as in Umar's investigation) it is important to use a flat stretch of road.

Other calculations from graphs

You might be asked to:

- calculate acceleration from a graph of velocity against time (see lessons 6.3 and 6.4)
- calculate resistance from a graph of voltage against current (see lesson 6.21).

Worked example

Calculate the gradient of the line on Umar's graph.

Step 1 Make the straight line into a triangle (as large as possible) by drawing a vertical line and a horizontal line. Use this triangle to find the change in y-axis and change in x-axis of the best-fit line.

Step 2 The vertical side of the triangle shows you the change in the y-axis. The bottom of the line is at 0 m and the top of the line is at 1800 m. So the change in y is 1800 m.

Step 3 The horizontal side of the triangle shows you the change in the x-axis. The start of the line is at 0 s and the end of the line is at 100 s. So the change in x is 100 s.

Step 4 The gradient $= \dfrac{\text{(change in y-axis)}}{\text{change in x-axis}} = \dfrac{1800}{100} = 18 \text{ m/s}$.

Activity B

1 m/s = 2.24 miles per hour.

1 Use this conversion factor to calculate the speed of the car, in miles per hour, in Umar's test.

2 Is the cruise control working properly?

Just checking

1 An electrician is testing the resistance of a bulb in a set of Christmas tree lights. She measures the current when different voltages are applied. Look at the graph of her results.

   (a) Predict the current when a voltage of 5 V is applied.

   (b) Calculate the resistance of the bulb.

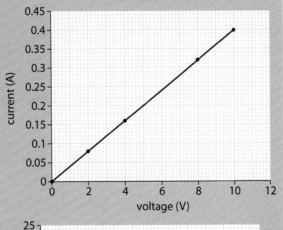

2 A rollercoaster engineer is trying to calculate the acceleration of a trolley in a model of a rollercoaster ride. He measures the velocity at different times in the ride. Look at the graph of his results.

   (a) Calculate the gradient of the line between 2 s and 8 s.

   (b) What is the acceleration during this time?

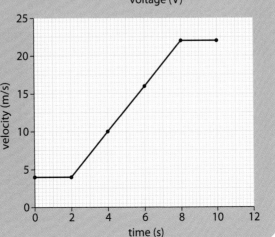

Lesson outcome

You should be able to carry out calculations using data obtained from a graph.

8.14 Anomalies on graphs

Get started

What are anomalous results? How do you recognise an anomalous result in a list of measurements?

Link

You learned about heat packs in lesson 5.4.

Link

You already know about anomalous results from lesson 8.6 and lesson 8.9.

Activity A

Explain why Kate thinks that the result at 40 cm³ does not fit the pattern in the rest of the data.

Kate is a product designer working on the design of a new heat pack based on a neutralisation reaction between ethanoic acid and sodium hydroxide. She wants to investigate which volume of the two chemicals produces the highest temperature rise.

In one part of her investigation she adds different volumes of sodium hydroxide to $50 \, cm^3$ of ethanoic acid and measures the temperature rise.

Identifying anomalies

The results of Kate's investigation are shown on this graph.

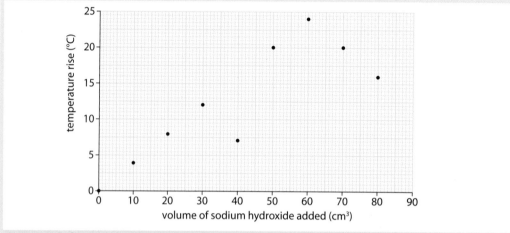

Kate can use her graph to identify an anomalous result.

Kate thinks there is an anomalous result in her data. How does she know?

Remember that an anomalous result is one that falls outside the expected range of measurements. In this case, the anomalous result does not fit the pattern of the other results.

Kate decides that the anomalous result is the temperature rise she measured when she added $40 \, cm^3$ of sodium hydroxide.

Explaining anomalous results

The most likely reason for an anomalous result is an error in the experimental process.
- Kate could have made a mistake in what she did. She might have added the wrong volume of sodium hydroxide.
- The measuring equipment may not be **calibrated** correctly. This would give an incorrect reading for the temperature rise.
- Kate may not have controlled some of the variables. Perhaps she did not stir the reaction mixture as much in one of the experiments as in the others.

Dealing with anomalous results

Kate can deal with the anomalous result on her graph in one of two ways.
- She can ignore the anomalous result, and draw a best-fit line that shows the pattern in the rest of the data.
- She can repeat the measurement at $40 \, cm^3$. If the new result is not anomalous she can plot this on the graph instead.

Link

You can read more about carbon nanospheres, such as Buckminsterfullerene, in lessons 5.15 and 5.16.

Case study

My name is Hans. I was a research chemist in the 1980s when the first nanochemical, Buckminsterfullerene, was discovered. In our laboratory, we were heating up carbon in a vacuum and measuring the size of the particles that were formed. We expected that most of them would contain just a few carbon atoms, and that any larger molecules would be far too unstable to exist for very long.

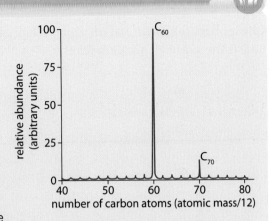

But when the results came back, there were massive amounts of a molecule with exactly 60 carbon atoms. At first we thought this was an anomalous result, but after we repeated the experiment many times we realised that something very exciting was happening, and later we realised that we had made a completely new stable form of carbon.

We often think that anomalous results must be bad news for an experiment, but here what we thought was an anomaly made us think again about how the rules of chemistry work, and helped our team to produce samples of this exciting new material.

Just checking

Rob is a hospital laboratory technician. He is checking the most effective dose of antibiotic to use against a bacterial infection. He adds three drops of antibiotic solution of different concentrations to plates on which bacteria are growing. Rob plots the graph below of diameter of the bacteria-free zone against the concentration of the solution.

1 Identify the anomalous result in Rob's data.

2 Explain your choice.

3 Suggest one possible cause of the anomalous result.

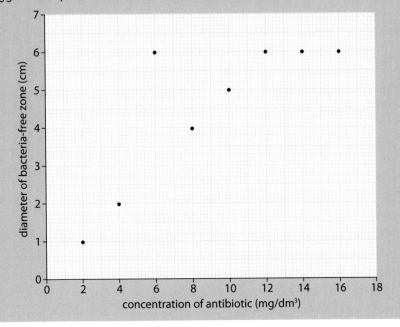

Lesson outcomes

You should be able to identify anomalous results in graphs and explain why they occur.

Get started

Try to explain to someone what happens to the current in an electrical circuit when you increase the voltage.

(a) Explain in words what happens.

(b) Try and sketch the pattern that you would see when you plot a graph of current against voltage.

Key terms

Conclusion – The answer to the question posed by the scientist's hypothesis at the start of an investigation.

Directly proportional – If the independent variable increases by a factor and this causes the dependent variable to increase by the same factor, then the two variables are directly proportional to each other.

Negative correlation – Data show a negative correlation if the dependent variable decreases as the independent variable increases.

Positive correlation – Data show a positive correlation if the dependent variable increases as the independent variable increases.

The SONAR system on this submarine must be tested regularly.

In the previous lesson you read about Kate's investigation to find which volumes of two reacting chemicals produce the highest temperature rise.

Kate repeats the investigation, using a higher concentration of sodium hydroxide. Here is the graph of Kate's results.

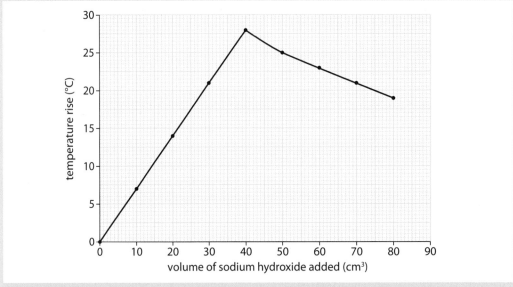

Kate's graph shows temperature rise when different volumes of sodium hydroxide are added to 50 cm^3 of ethanoic acid.

Kate looks at the graph and sees that the maximum temperature rise occurs when 40 cm^3 of sodium hydroxide is added. This is the **conclusion** of Kate's investigation. It is the answer to the question that she asked herself at the beginning of the investigation.

But very often a conclusion is not as simple as Kate's.

Richard is a SONAR operator for a submarine. He is testing the SONAR system on his submarine. He knows the depth of the ocean floor at several locations, and he measures the time taken for the SONAR signal to return to the sensor. Here is a graph of his results.

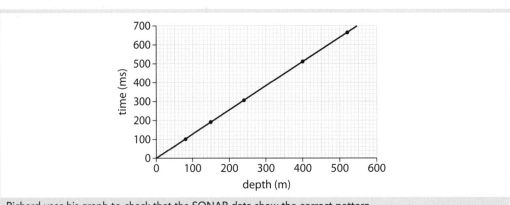

Richard uses his graph to check that the SONAR data show the correct pattern.

Scientists describe patterns in data by using a number of important terms. Richard's graph shows a **positive correlation** because as the depth increases the time taken for the signal to return also increases.

Richard checks to see if the two variables are **directly proportional**. This is a special case of positive correlation.

Richard's hypothesis was that if the depth doubles the time will also double. He could check this using data from the graph. But Richard knows that if two variables are proportional to each other then the graph will be a straight line passing through zero on both axes. He looks at the graph and sees that this is the case.
Richard's full conclusion is as follows:

- there is a positive correlation between depth and time
- time is directly proportional to depth.

Look again at Kate's graph showing how temperature rise depends on the volume of sodium hydroxide added. Between $40\,cm^3$ and $80\,cm^3$ the temperature change goes down as the volume of sodium hydroxide increases. This is a **negative correlation**.

Kate checks whether the two variables are **inversely proportional** between $40\,cm^3$ and $80\,cm^3$. If they are then the temperature rise should halve as the volume of sodium hydroxide doubles.

Kate decides that the two variables are not inversely proportional to each other above $40\,cm^3$.

Activity A

Look at the graph of Richard's results.

1 Find the time taken by the signal at a depth of
  (a) 200 m
  (b) 400 m.

2 Explain how this confirms that time is directly proportional to depth.

Activity B

1 In Kate's investigation, what is the temperature change for:
  (a) $40\,cm^3$ of sodium hydroxide?
  (b) $80\,cm^3$ of sodium hydroxide?

2 Explain how these values confirm that the two variables are not inversely proportional to each other.

3 What is the pattern in Kate's data between 0 and $40\,cm^3$ of sodium hydroxide? Use your answer, and the other information on this page, to write a full description of the overall pattern in Kate's data.

Just checking

Look at these four graphs, labelled **1 – 4**. They show four different patterns in data.

Decide which description **(A – D)** correctly matches each graph:

(A) positive correlation, not directly proportional

(B) negative correlation, not inversely proportional

(C) positive correlation and directly proportional

(D) negative correlation and inversely proportional.

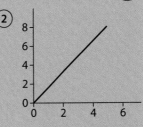
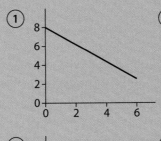
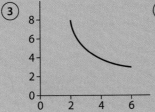
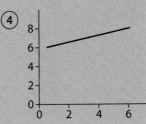

Lesson outcome

You should be able to describe the trends and patterns in data.

Key term

Inference – A deduction that can be made from a scientific conclusion. This can lead to further investigations to come to a new conclusion.

Link

This lesson links with the information about blood groups in lesson 7.15.

Sayeed is a forensic scientist. He is testing some blood that was found at a crime scene. He also tests the blood of three suspects. He is trying to find out the blood group of each of the samples.

Conclusions and inferences

The table below shows Sayeed's results. He has not yet drawn a conclusion from the data.

The hypothesis that Sayeed is trying to prove in his investigation is:

'The blood group of the sample from the scene can be used to prove that certain suspects could not have left the blood sample at the scene.'

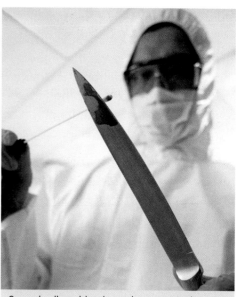

Sayeed collects blood samples to test in the forensic laboratory.

His conclusion is just the answer to the question posed by the hypothesis.

Blood sample	Antigen A	Antigen B	Blood group
From crime scene	✓	✗	
Suspect A	✓	✓	
Suspect B	✓	✗	
Suspect C	✗	✗	

Sayeed explains his conclusion to a detective. His conclusion is that the blood could not have come from suspect A or C, but it could have come from suspect B.

This is an **inference** that has been made using the evidence of a scientific conclusion. Sayeed will need further data to come to a conclusion that can be put before a court, as other people also have this blood group. For example, if suspect B could prove using CCTV that he was 100 miles away at the time of the crime, his blood group would be irrelevant.

Activity A

1 Use information from lesson 7.15 to write down the blood group of each of the four samples.

2 Explain why the detective cannot say that the blood at the crime scene definitely comes from suspect B.

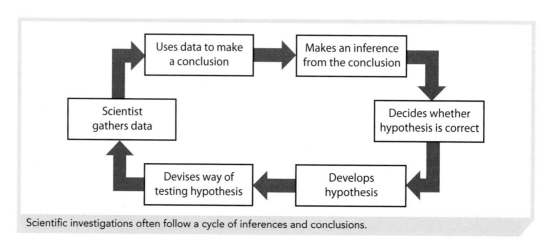

Scientific investigations often follow a cycle of inferences and conclusions.

Evaluating tests

In Sayeed's forensic investigation, he can be very sure about his conclusion because the tests produce results that are *repeatable*. This means that if Sayeed repeats the test, it will always give the same result.

Not all tests are repeatable, which means that scientists cannot always be so certain about their conclusion.

Kelly is a food scientist working for a chocolate manufacturer. She is comparing the sweetness of the company's products.

She asks six people to rate the product from 1 (not sweet at all) to 5 (very sweet). The results are shown in the table.

Product	Average score
Milki	3.6
Smoothi	4.7
Nutti	2.7
Toffi	4.9

Kelly concludes that Toffi is the sweetest product. Kelly's boss says that he can't be very sure about the conclusion because the measurements on which it is based are not very repeatable.

Kelly decides to use a practical method to measure the sweetness, as she thinks this will produce better quality evidence. She takes a small sample of chocolate and dissolves it in hot water. Then, she uses a test strip that turns different colours depending on the concentration of the sugar glucose.

Did you know?

Glucose test strips are used by people with diabetes to monitor their blood sugar level.

Kelly repeats the experiment three times and takes an average of the concordant results. Although these measurements are now more repeatable, Kelly must also think about whether the test is *valid*. A test is valid if:

- it is fair, i.e. all the important variables are controlled
- the method is not biased. Biased methods are ones that always produce results that are too low or too high.

You can get practice in applying ideas from this lesson in the 'Just checking' feature in lesson 8.17.

Discussion point

Suggest reasons why Kelly's conclusion is not very repeatable.

Activity B

1. List the things that Kelly should do to make her test fair.

2. Suggest a reason why Kelly's experiment using the glucose test strip might be biased. Hint: the test strip gives a reading for the concentration of glucose only.

Lesson outcomes

You should be able to draw inferences from a conclusion and evaluate the method that is used to produce the conclusion.

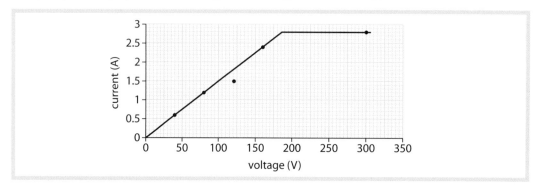

Get started

What is a hypothesis? What do you think is meant by 'evaluating a hypothesis'?

Does this filament light bulb obey Ohm's law?

Activity A

Use ideas from lesson 6.22 to explain why the resistance of a filament bulb increases at high currents.

Did you know?

William Coolidge, a US physicist, engineer and inventor, discovered how to make tungsten ductile, greatly increasing the life of the tungsten filaments in electric light bulbs.

William Coolidge, 1873–1975.

Dwayne is a lighting engineer. He needs to find out what happens to the resistance of a powerful filament light bulb.

His hypothesis is that the bulb will obey Ohm's law when the current is low, but at high currents the resistance will increase.

Dwayne plots a graph of current against voltage to test his hypothesis.

Dwayne looks at the graph of his results and decides that there is some evidence to show that his hypothesis is correct.

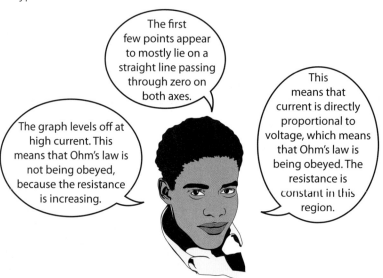

The first few points appear to mostly lie on a straight line passing through zero on both axes.

This means that current is directly proportional to voltage, which means that Ohm's law is being obeyed. The resistance is constant in this region.

The graph levels off at high current. This means that Ohm's law is not being obeyed, because the resistance is increasing.

Dwayne shows the graph to Colleen, one of his colleagues. She has some doubts.

There could be other ways of drawing a best-fit line.

The only evidence for the graph flattening off is the final point. This could be an anomaly, so there isn't enough evidence for the resistance increasing.

Only three points actually lie on the straight line, because one seems to be an anomaly. This isn't really enough evidence to be sure that the two variables are proportional to each other.

Activity B

1 On a copy of Dwayne's graph, try drawing some other best-fit lines to show other possible patterns in the data:

(a) draw a best-fit line that levels off more gradually than Dwayne's line

(b) draw a best-fit curve by ignoring the point at 160 V.

2 Which of the three best-fit lines do you think is the most likely to be the true pattern?

Improving the evidence

Dwayne realises that he does not have any information for voltages between 160 V and 300 V. He decides to investigate at least two more voltages in this range. He also re-investigates the anomaly at 120 V.

This will help him to be more certain about the best-fit line. He will then have better evidence from which to draw a conclusion that his hypothesis is correct.

In the previous lesson, you read about ways of making the data more reliable and more valid. You should:

- choose a way of measuring that gives data that are repeatable
- repeat the measurements and take an average of concordant values
- make sure that the experiment is a fair test
- avoid any bias in the experiment.

Just checking

Melissa is a chemist working for an oil refining company investigating the energy given out by different liquid fuels when they are burnt. She tests the hypothesis that the longer the chain of carbon atoms, the less energy is given out per gram of fuel.

She burns some fuel in a spirit burner and uses the energy it gives out to heat 100 g of water in a beaker. Here are her results.

Fuel	Number of carbon atoms	Mass burned (g)	Energy given out (kJ)
Hexane	6	1.0	24.2
Heptane	7	1.1	26.5
Dodecane	12	0.9	21.4

1 How well do these results support Melissa's hypothesis?

2 Suggest some changes Melissa could make to her investigation so that she is more certain about whether the hypothesis is correct.

Assessment Zone

How you will be assessed

You will take a written test on the same day and time as other learners – your teacher will let you know the date of the test. The test will last 1 hour and 15 minutes and you should aim to answer all the questions on the paper. For some questions you will need to use a calculator.

Your BTEC written test will contain these types of questions:

- **multiple-choice questions** where the answers are available and you have to choose the answer(s) that fit(s)
- **short-answer questions** where you are asked to give a short answer worth 1–2 marks
- **long-answer questions** where you are asked to give a longer answer which could be worth up to 6 marks.

Test tips

- At the start make sure you have read the instructions, that you can see the clock and that you are comfortable to write.
- Watch the time – you should aim to spend about a 'minute per mark'. Some early questions will take less time than this and some later ones more.
- If you get stuck on a question move onto the next one and come back to that question at the end.
- The space given for your answer will show you the type of answer required, e.g. if two answers are required you may see the answer space divided up for two answers.
- Remember that you can use more paper if necessary, e.g. because you have made a mistake or you need more space for your answer.
- Plan your longer answers – read the question carefully and think about the key points you will make. How well you structure your argument will be important.

Please note: the questions on the following pages are for practice. They do not represent a real test paper.

Assessment practice

Here are some questions for assessing scientific skills. Tips are given to help you understand what the question is asking and how to answer it. Full answers to the questions have also been included.

Question 1

Richu is investigating the connection between fitness level and age. She has a hypothesis that the older a person is, the less fit they are.

She asks three people of different ages to exercise vigorously for 1 minute and then measures their pulse rate.

Here are her results.

Person	Age	Pulse rate (beats per minute)
Jane	26	106
Sally	46	123
Rob	59	148

You may need to make some inferences from the data – for example to think about the link between pulse rate and fitness. You need to state the link and explain why Richu might think this means the results support the hypothesis.

(a) Richu decides that these results support her hypothesis.
Explain why Richu might think this. [2]

The older the person the higher the pulse rate, which could indicate that the person is more unfit.

(b) Richu's teacher tells her that her hypothesis is not strongly supported by the evidence because of the way she carried out the experiment.

(i) Give three possible reasons for this. [3]

You need to look at any information given in the question. This will help you decide whether the experiment was a fair test and whether the method is a repeatable way of testing fitness.

She did not keep all other variables constant – for example two of the people in the experiment were women and one was a man. This means that the comparisons are not relevant.

She only studied three people, which is too small a sample to be confident that there were no anomalies.

She did not measure the resting pulse rate.

(ii) Explain one way in which Richu could improve her experiment to make her conclusion more valid. [2]

This means that you need to do more than just say what she should do – you need to explain how it helps to strengthen the evidence for her conclusion.

She could measure the difference between the resting pulse rate and the pulse rate after exercising because this is a more valid way of determining fitness than just measuring the pulse rate after exercise.

Question 2

Robin carried out an investigation into the motion of a trolley. He measured the distance travelled by the trolley at different time points.

He repeated the experiment three times.

Time (s)	Distance travelled (Run 1) (m)	Distance travelled (Run 2) (m)	Distance travelled (Run 3) (m)	Mean distance (m)
2	0.32	0.33	0.31	0.32
4	0.64	0.64	0.67	0.65
6	0.96	0.94	0.95	0.95
8	1.24	1.23	1.22	1.23
10	1.55	1.28	1.57	1.56

To explain your answer you should use your understanding of the term 'anomalous'. You need to state which value and explain why.

This question tests that you can calculate a mean and that you know how to handle anomalous data. One of the results at 10 s is anomalous so you should not include it in the calculation.

Your graph needs to be clear, so make it fill as much of the available space as possible. You must decide on a suitable scale and axes, label the axes and plot the points correctly.

A best-fit line can mean a straight line or a curve, but here the points seem to lie close to a straight line. The best-fit line doesn't need to go through all the points but should go as close as possible to all points, except any that are anomalous.

(a) **(i)** The data for 10 s include an anomalous result. Identify this result and explain how you know it is anomalous. [2]

The 1.28 m result is anomalous because it does not fit the pattern of results for the other runs and times.

(ii) Calculate an appropriate value for the mean distance travelled after 10 s. Write your answer in the table. [2]

(b) **(i)** Plot a graph of the average distance against time.

Use the graph paper below. Draw a best-fit line to show the pattern in the data. [2]

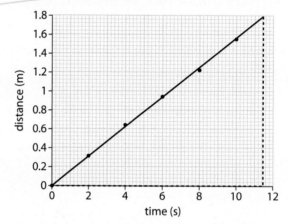

(ii) Robin concludes that the trolley is travelling at a constant speed throughout the experiment. How can he tell this from the graph? [1]

The graph of distance against time has the same slope throughout the experiment.

(iii) The speed of a vehicle is given by the gradient of a graph of distance against time.

Calculate the average speed of the vehicle in Robin's investigation. Give your answer to 3 significant figures.

Change in y-axis = 1.80   Change in x-axis = 11.5
So gradient = 1.8 / 11.5 = 0.157 m/s [3]

The first line of the question tells you how to do this – by calculating the gradient of the graph. You should draw lines on the graph to make the best-fit line into a triangle.

You will use the graph to answer this question, but you will also need to use some understanding of the subject of motion which you studied in Unit 6.

How to improve your answers

Here is a question that assesses part of learning aim A:

'Understand how to produce a good plan for an investigation'.

Harry is investigating the chemical reaction between zinc and hydrochloric acid. He knows that when zinc is added to hydrochloric acid, the solution gets hotter.

He thinks the temperature rise will be proportional to the amount of zinc added. Write a plan to test this hypothesis. [6]

Read the three learner answers below, together with the feedback.

This activity will help you learn more about how to answer longer answer questions.

Try to use what you learn here when you answer questions like this in your test.

The plan needs to include details of all the equipment needed.

It's important to say the same volume and same concentration of acid.

Learner 1

Add zinc to hydrochloric acid in a plastic cup and measure how hot it gets. Then do this again with more zinc but the same hydrochloric acid.

Feedback:

This answer might pick up 1 or 2 marks. One piece of equipment is mentioned and the idea about using different amounts of zinc is described. However, there are several other pieces of equipment which should also be listed. There is no mention of how many different masses of zinc to use or any idea of a suitable range, nor any mention of amount or concentration of the acid. The candidate has made an attempt to control one of the variables but hasn't described this very clearly. .

The answer doesn't explain what equipment is used to measure these.

This links in with the hypothesis, but doesn't explain how to test a proportional relationship.

The concentration of the acid must also be kept the same.

Learner 2

Measure out 100 cm³ of hydrochloric acid and 1g of zinc. Add them together in a plastic cup and use a thermometer to measure the temperature rise. Then use 2g, 3g, 4g of zinc and see what happens. Keep the hydrochloric acid at 100 cm³. The temperature should go up.

Feedback:

This would probably pick up 4 of the 6 marks available. The method is written in a logical order and another learner could probably follow the instructions. Not all the equipment is mentioned – some of the measuring instruments aren't listed. The concentration of the acid should be stated and the starting temperature should be recorded. The range of masses is fine, but the answer doesn't explain how to test the quantitative hypothesis.

Learner 3

All the measuring equipment is listed.

There is a lot of fine detail, making it easy to follow the instructions.

The answer describes very clearly how to test the hypothesis.

> Measure 1g of zinc on a balance (to 2 decimal places) and 100 cm³ of hydrochloric acid in a measuring cylinder. Add the acid to a plastic cup. Measure the temperature of the acid using a thermometer (0-100°C) before the reaction, then add the zinc and stir it. Measure how hot the solution gets. Then repeat using the same volume and concentration of acid and the same starting temperature. Use five different masses of zinc: 1g, 2g, 3g, 4g, 5g. If the temperature rise doubles when the mass doubles then they are proportional.

Feedback:

This is an excellent answer which covers all the required points and will score 6 marks. All the relevant equipment is listed and the instructions are very detailed and in a logical order. The range and number of masses used are excellent and there is a very clear description of how to test the hypothesis. An alternative way would be to plot the data on a graph and see if the best-fit line is a straight line passing through zero on both axes.

Test tips for this question

- You need to know the full meaning of scientific words, such as 'proportional' and 'hypothesis'.
- This question asks you to write a plan to test a hypothesis and there are 6 marks available for doing this. So you will need to make several different points and organise them into the correct order.
- You will need to
  - choose equipment
  - identify the variables which you would change and any that you need to control
  - select a suitable range and number of measurements to make
  - order your method in a logical way
  - explain how the method would be used to test the hypothesis.
- Try to avoid using words such as 'it' or 'they' in your answers because it may not be clear what you are referring to.
- Make sure that you use correct scientific words in your answers, like the full names of any chemical substances or, in this case, words like 'variable' or 'control'.

Assess yourself

Question 1

Antibiotics are substances which kill bacteria. Jess and Mohamed are going to investigate how the concentration of an antibiotic affects its effectiveness. They soak some discs of filter paper in antibiotic solutions of different concentrations. They place each of the discs on a separate Petri dish on which a colony of bacteria is growing. After 24 hours they measure the diameter of the zone where no bacteria are growing.

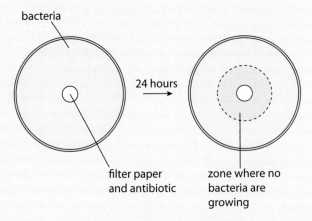

bacteria

24 hours

filter paper and antibiotic

zone where no bacteria are growing

(a) (i) The independent variable in the students' experiment is the concentration of the antibiotic. Explain how you can tell this from the information above. [1]

(ii) What is the dependent variable in the students' experiment? [1]

(b) The highest concentration of antibiotic which the students used was 100 mg/dm³. Suggest other concentrations which the students might use in their investigation. [3]

(c) Handling microorganisms can be hazardous.

(i) Describe one risk of handling microorganisms. [1]

(ii) Suggest one precaution the students should take during their investigation to minimise this risk. [1]

Question 2

A team of health professionals investigated the lifestyle of a group of 60 people. The team used a survey to obtain data about weight, smoking and amount of exercise taken by individuals in the group. The table below shows the results of the surveys.

Weight	Below healthy weight	Healthy weight	Above healthy weight
Number of individuals	2	42	16
Smoking behaviour	Non-smoker	Occasional smoker	Regular smoker
Number of individuals	26	5	29
Amount of exercise per week	Less than 1 hour	Between 1 and 3 hours	Above 3 hours
Number of individuals	38	12	9

The team present their findings using different types of graph.

(a) On a piece of graph paper, draw a bar-chart to represent the data about exercise. [3]

(b) Copy and complete the pie chart below to show the data about smoking. [2]

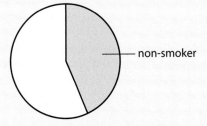

non-smoker

(c) One of the team plans to use a line graph with a line of best-fit to display the data about weight. Give one reason why this would not be a sensible way to display these data. [1]

Question 3

A drug company has developed a new analgesic (pain-killing drug). They test it on 100 patients who are in pain. They give the patients a dose of the drug and after 15 minutes ask the patients if their pain is better, worse or the same as before taking the drug. The drug company also do the same test on the same group of patients using a very common painkiller, paracetamol. The table below shows the results of the tests.

	Number of patients reporting		
Drug	pain is better	pain is worse	pain is the same
Paracetamol	53	3	44
New analgesic	69	6	25

(a) What can the company conclude from the results of these tests? [1]

(b) (i) Discuss how well the evidence supports the conclusion. [3]

    (ii) Suggest one way in which the drug company could improve or extend this investigation to give them better evidence for how effective their new drug is. [1]

Question 4

Chris and Vicky are investigating the reaction between magnesium ribbon and hydrochloric acid. They know that this is an exothermic reaction. They decide to measure the temperature change of the hydrochloric acid when different masses of magnesium ribbon are added to the same volume and concentration of hydrochloric acid.

(a) Write down a hypothesis to state what you expect to happen to the temperature change when they add a greater mass of magnesium. [2]

(b) Use relevant scientific ideas to justify your hypothesis. [2]

Question 5

Nick works in the quality control department of a chemical firm. He monitors the purity of products which are being made in pilot plants. One pilot plant is making ethanoic acid. Nick takes samples of ethanoic acid at different times during the production process and tests the purity of each sample. The table below shows the results of Nick's tests.

Time sample taken (hours)	1	4	8	12	16	20
Purity (%)	10	40	68	90	96	100

(a) On a piece of graph paper, plot a line graph of these results. Draw a best-fit line to show the pattern in the data. [3]

(b) (i) Describe the pattern in the data. [3]

    (ii) Use the graph to estimate a value for the purity of the sample after 10 hours. [1]

Question 6

Tom and Gemma want to investigate the energy released when different fuels are burnt. They use the apparatus shown.

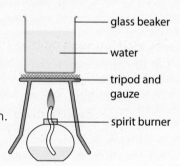

- glass beaker
- water
- tripod and gauze
- spirit burner

They use a thermometer to take the temperature of the water at the start of the experiment. They then light the burner and use it to heat the water. After some time they take the new temperature of the water.

(a) Identify two things that Tom and Gemma should do to make their experiment a fair test. [2]

(b) Suggest two other improvements that they could make to the method or the apparatus to improve the quality of the evidence obtained in their investigation. [2]

Question 7

Jade and Ben are investigating the average speed of a trolley as it rolls down a slope.

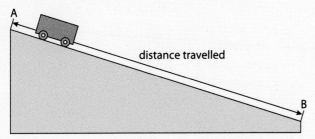

They measure the distance between two points, A and B, on the slope and use a stop-clock to time how long the trolley takes to travel from A to B. Jade and Ben repeat the experiment several times. The table below gives their results.

Experiment	Time taken to travel from A to B (seconds)
1	1.9
2	2.0
3	3.4
4	1.8

Distance between point A and point B = 2.6 m.

(a) (i) The result from experiment 3 was an anomaly. How can you tell? [1]

   (ii) Suggest one thing that could have happened in experiment 3 to cause the anomaly. [1]

(b) Calculate an appropriate mean value for the time taken for the trolley to travel from A to B. [1]

(c) In another experiment, the time taken to travel 3.2 m was 2.5 seconds.

   The average speed of an object is given by the equation:

   $$\text{average speed} = \frac{\text{distance travelled}}{\text{time taken}}$$

   Calculate the average speed of the trolley in this experiment. Give your answer to an appropriate number of significant figures. [2]

Question 8

Lizzie and Bradley are carrying out an investigation to measure the speed of a ball-bearing when it is dropped from different heights. They plot a graph of speed against height.

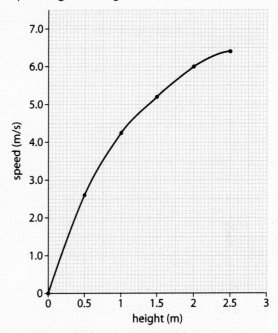

(a) Look at the four descriptions below. Which statement correctly describes the pattern in the data? [1]

   (1) The two variables are directly proportional to each other.

   (2) There is a negative correlation between the two variables.

   (3) The two variables are inversely proportional to each other.

   (4) There is a positive correlation between the two variables.

(b) Lizzie and Bradley use the data to calculate the kinetic energy (KE) of the ball-bearing which was dropped from a height of 2 m.

   They find that the KE of the ball-bearing as it hits the ground = 0.5 J.

   They discover from research that the KE of an object is directly proportional to the height which it is dropped from.

   What would the KE of the ball-bearing be if it was dropped from a height of 4 m? [1]

Question 9

Laura is investigating what happens to the resistance of a light-dependent resistor (LDR) when she shines lights of different brightness on to it. She places the LDR in a simple circuit.

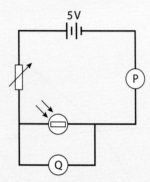

(a) Name the two different meters labelled P and Q on this diagram. [2]

(b) Write a plan that Laura could follow to carry out her investigation. [5]

The periodic table of elements

The periodic table of elements

Key

relative atomic mass
atomic symbol
name
atomic (proton) number

1	2											3	4	5	6	7	0
							1 **H** hydrogen 1										4 **He** helium 2
7 **Li** lithium 3	9 **Be** beryllium 4											11 **B** boron 5	12 **C** carbon 6	14 **N** nitrogen 7	16 **O** oxygen 8	19 **F** fluorine 9	20 **Ne** neon 10
23 **Na** sodium 11	24 **Mg** magnesium 12											27 **Al** aluminium 13	28 **Si** silicon 14	31 **P** phosphorus 15	32 **S** sulfur 16	35.5 **Cl** chlorine 17	40 **Ar** argon 18
39 **K** potassium 19	40 **Ca** calcium 20	45 **Sc** scandium 21	48 **Ti** titanium 22	51 **V** vanadium 23	52 **Cr** chromium 24	55 **Mn** manganese 25	56 **Fe** iron 26	59 **Co** cobalt 27	59 **Ni** nickel 28	63.5 **Cu** copper 29	65 **Zn** zinc 30	70 **Ga** gallium 31	73 **Ge** germanium 32	75 **As** arsenic 33	79 **Se** selenium 34	80 **Br** bromine 35	84 **Kr** krypton 36
85 **Rb** rubidium 37	88 **Sr** strontium 38	89 **Y** yttrium 39	91 **Zr** zirconium 40	93 **Nb** niobium 41	96 **Mo** molybdenum 42	[98] **Tc** technetium 43	101 **Ru** ruthenium 44	103 **Rh** rhodium 45	106 **Pd** palladium 46	108 **Ag** silver 47	112 **Cd** cadmium 48	115 **In** indium 49	119 **Sn** tin 50	122 **Sb** antimony 51	128 **Te** tellurium 52	127 **I** iodine 53	131 **Xe** xenon 54
133 **Cs** caesium 55	137 **Ba** barium 56	139 **La*** lanthanum 57	178 **Hf** hafnium 72	181 **Ta** tantalum 73	184 **W** tungsten 74	186 **Re** rhenium 75	190 **Os** osmium 76	192 **Ir** iridium 77	195 **Pt** platinum 78	197 **Au** gold 79	201 **Hg** mercury 80	204 **Tl** thallium 81	207 **Pb** lead 82	209 **Bi** bismuth 83	[209] **Po** polonium 84	[210] **At** astatine 85	[222] **Rn** radon 86
[223] **Fr** francium 87	[226] **Ra** radium 88	[227] **Ac*** actinium 89															

* The lanthanoids (atomic numbers 58–71) and the actinoids (atomic numbers 90–103) have been omitted.

The relative atomic masses of copper and chlorine have not been rounded to the nearest whole number.
For radioactive isotopes the relative atomic mass number is given in square brackets.

Glossary

acceleration – The change in velocity divided by the time taken.

accurate – An accurate measurement is close to the true value.

acquired immunity – Immunity acquired after an infection or a vaccination has stimulated a specific immune response and some B cells have developed into memory cells that stay in your body.

action-reaction pair – A pair of forces which are equal and opposite; they are the same type of force, but act on different bodies.

addition reaction – A reaction in which two molecules join together to make a single product molecule.

air resistance – The force opposing the motion of an object moving through the air, sometimes called drag.

alcohols – Organic molecules that contain the functional group -OH.

alkanes – Hydrocarbons which contain C-C and C-H single bonds only.

alkenes – Hydrocarbons which contain one or more C=C double bonds.

ammeter – An electrical instrument that measures current.

amniocentesis – A diagnostic test carried out during pregnancy to find out whether the fetus could develop, or has developed, an abnormality or serious health condition. A needle is used to extract a sample of amniotic fluid which can be examined and tested for a number of conditions.

analgesic – A drug used to reduce or relieve pain.

aneurysm – A balloon-like bulge in an artery which, when it grows large, can burst; this can be fatal.

angle of incidence (incident ray) – The angle between the incident ray (the ray of light arriving at a mirror) and the normal.

angle of reflection (reflected ray) – The angle between the reflected ray and the normal.

anomalous (anomaly) – A result is anomalous if it falls outside the normal (or expected) range of measurements.

anorexia – A psychological eating disorder in which the sufferer does not eat or eats very little.

antenatal – Before birth.

antibacterial – A type of antibiotic that only kills bacteria.

antibiotics – Drugs that can destroy or inhibit the growth of microorganisms such as bacteria and some fungi, but not viruses.

antibodies – Proteins produced in the blood that attack and kill pathogens such as bacteria or viruses.

anti-fungal – A drug that kills or slows the growth of fungi.

antigens – Chemicals, usually proteins, on the surface membranes of cells that can activate B and T lymphocytes to carry out a specific immune response.

antihistamine – A substance that reduces the itching and swelling that histamine produces.

antiviral – A drug used to treat infections caused by viruses, such as flu or measles.

applied force – A force that is applied to an object by a person or by another object.

atherosclerosis – A disease caused by the build-up of fatty deposits on the walls of arteries.

balanced forces – When forces are balanced, the resultant force is zero and the object remains at a constant speed.

bar chart – A diagram showing statistical information. Bars, of equal width, represent frequencies. The lengths of the bars are proportional to the frequencies. Sometimes called a bar graph.

best-fit line – A line drawn on a scatter graph to represent the best estimate of the relationship between the variables.

biased (source) – A statement or article that deliberately favours one side of an argument over another.

biodegradable – A substance that can be broken down by the action of living organisms such as bacteria.

biopsy – Taking a small sample of tissue to test for diseases and other problems.

body mass index (BMI) – A number calculated from a person's weight and height that provides a reliable indicator of body fatness for most people.

braking distance – The distance travelled by a vehicle while the brakes are working to bring it to a halt.

calibrated – Set against a known scale.

calliper – A device for taking very small measurements.

calorimeter – An apparatus for measuring the energy change during a reaction.

cancer – A group of diseases caused by uncontrolled cell division leading to the formation of tumours or growths.

carboxylic acids – Organic molecules that contain the functional group -COOH.

cardiovascular – Relating to the heart and blood vessels.

catalyst – A substance that speeds up a chemical reaction but is not itself used up, so it can be used again.

cholesterol – A fat which is made in the liver and carried around the body in the blood. High levels are associated with an increased risk of heart disease.

chorionic villus sampling – A diagnostic test carried out during pregnancy to detect specific abnormalities in an unborn baby. Cells are taken from the placenta and tested for genetic defects.

chromosome – A long thread of a molecule called DNA. Each chromosome has a series of genes along its length.

cilia – The small hairs on the surface of some cells.

composite – A material made up of two or more different materials, often with differing physical or chemical properties.

compression – Area where molecules are closer together. This will be an area of high pressure.

compressive force – A force that compresses or squashes a material.

concave – Thinner in the middle than at the edges.

conclusion – The answer to the question posed by the scientist's hypothesis at the start of an investigation.

concordant – In agreement. Results are concordant if they fall within a narrow range.

conductor – A material that can conduct heat and electricity easily. Metals are good conductors.

conservation of energy – The principle that energy cannot be created or destroyed, only changed from one form to another, or transferred from one object to another.

constant – A number or quantity that does not vary.

control variable – A variable that must be kept constant during the investigation if it is to be a fair test.

converge – Come together.

convex – Thicker in the middle than at the edges.

critical angle – The angle of incidence at which total internal reflection, rather than refraction, begins to occur.

cystic fibrosis (CF) – A genetic disorder caused by inheriting two copies of a recessive allele. It causes thick mucus to collect in the lungs, making breathing difficult, and also stops food being digested properly.

data – Numerical information.

data logger – An electronic device which records data.

decompose – To break down a compound into simpler substances.

deficiency – Lack of an essential nutrient in the diet.

deform – Change shape.

degenerative condition – A disease where an organ becomes progressively damaged with loss of function.

dependent variable – A variable that is measured for each and every change in the independent variable.

diminished image – An image that is smaller than the object.

directly proportional – If the independent variable increases by a factor and this causes the dependent variable to increase by the same factor, then the two variables are directly proportional to each other.

dispersion – The separating of the colours in light, for example when white light passes through a prism.

displacement – The shortest distance from the initial to the final position of an object that has moved.

displayed formula – A diagram showing every atom and every bond in the molecule.

dissolve – When a substance splits up and mixes with a liquid to make a solution.

donor – Person giving; in this context, giving blood or a body organ.

Down syndrome – A condition caused when an individual has three copies of chromosome 21; it can affect physical and mental development.

drag force – Resistance to motion through a medium e.g. air or water.

drug – A substance that alters the body's metabolism.

ductile – Can be stretched into wires.

echo – The repetition of a sound produced when the sound is reflected back from a surface.

elastic – A substance that will return to its original shape and size after it has been stretched or squashed.

endorphins – Chemicals produced during exercise which make you feel good and reduce stress levels.

endothermic – A reaction that takes in energy (in the form of heat) from the surroundings, causing a fall in temperature.

energy – Whenever something happens, energy is involved. It can exist in many forms but is always measured in joules (J).

energy profile diagram – A way of displaying the changes in enthalpy (energy) which happen during a chemical reaction.

energy transformation diagram – A diagram where energy flow is shown by arrows.

enthalpy change – The amount of heat transferred to or from the chemicals in a reaction.

esters – Organic molecules that contain the functional group -COO-.

ethical issue – A problem for discussion that concerns whether a particular action or decision is morally right or wrong.

eukaryotic cells – Cells that contain a nucleus.

exothermic – A reaction that gives out energy (in the form of heat) to the surroundings, causing a rise in temperature.

fair test – (of a hypothesis) A test in which the independent variable is the only thing allowed to change during the investigation.

feedstock – A substance used as the starting material for an industrial reaction.

fermentation – When microorganisms break down large molecules to produce different substances, including foodstuffs and drugs; e.g. using yeast to break down sugars to produce ethanol.

filament bulb – A bulb whose resistance increases when it gets hot.

flammable – Can easily catch fire.

focal length (f) – The distance from the centre of a lens to its focal point.

focal point (F) – For a convex lens, the point at which the refracted light rays converge. For a concave lens, the imaginary point where the refracted rays of light appear to come from.

force – An interaction on an object that can cause it to accelerate.

fractional distillation – A method used to separate crude oil into different fractions depending on their boiling points.

fractions – The different component parts of crude oil.

frequency – The number of cycles in one second.

friction – The force between two surfaces that resists motion.

fullerene – A molecule made up of at least 60 carbon atoms linked together in rings to form a hollow sphere or tube.

gradient – How steep a straight line is. A steeper line has a larger gradient.

graphite – A form of carbon that is soft and conducts electricity. It is used as a lubricant as the layers slip over each other easily.

gravitational potential energy (GPE) – The energy stored in an object because of its position in a gravitational field.

hazard – Something with the potential to cause harm.

health screening – Testing of a specific group of the population, before they show any symptoms of a disease. The aim is to find those who are in the early stages of the disease, or who are likely to develop the disease, so that treatment can begin earlier, giving a better outcome. In other words, early diagnosis leads to a better prognosis.

herd immunity – If everyone in a community is vaccinated against a disease during childhood then a pathogen cannot infect anyone and cannot spread.

histamine – A chemical released in the body during an allergic reaction; it triggers inflammation.

hydrocarbons – Chemicals that contain only hydrogen and carbon atoms.

hypothesis – A statement or prediction which can be tested by scientific investigation. The plural of hypothesis is hypotheses.

immunity – Having enough defences to avoid infection, disease or other unwanted invasion of the body.

immunosuppression – Suppression of the immune system.

incident ray – The ray of light arriving at a mirror.

independent variable – A variable which is purposely changed during an investigation.

inelastic – A substance that will not return to its original shape and size after it has been stretched or squashed. The change to the shape will be permanent.

infection – A disease caused by infecting agents, such as bacteria, fungi or viruses, which enter your body and make you ill.

inference – An inference is a deduction that can be made from a scientific conclusion. This can lead to further investigations to come to a new conclusion.

inflammation – Part of the non-specific immune response, involves swelling and pain.

innate immunity – The general (non-specific) immunity given by inflammation and phagocytosis; the first line of defence when pathogens enter your body.

insoluble – Unable to dissolve in a particular solvent, usually water.

insulator – A material that is a poor conductor of heat and electricity.

inversely proportional – If the independent variable increases by a factor and this causes the dependent variable to decrease by the same factor, then the two variables are inversely proportional to each other.

inverted image – An image that is upside down.

IVF – In vitro fertilisation; treatment for infertility. Eggs and sperm are mixed in a dish and two of the resulting embryos are placed into the mother's uterus.

joule – The joule (symbol J) is the scientific unit used to measure energy.

karyotype – A picture of an individual's chromosomes, typically arranged from largest to smallest; it can be used to identify genetic abnormalities.

Kelvin (K) – The unit in the Kelvin temperature scale. 1 K is the same temperature interval as 1 °C.

kinetic energy (KE) – The energy an object has because it is moving.

light dependent resistor (LDR) – A component that has high resistance in the dark and low resistance in the light.

light gates – Devices used with a data logger for measuring speed.

line graph – A graph consisting of a single straight line or of pieces (segments) of straight lines.

lubrication – Something placed between two moving surfaces to reduce the friction between them.

lymphocyte – A type of white blood cell that produces antibodies.

lysozyme – An enzyme that can break down the cell walls of bacteria and destroy them. It is found in blood, tears, mucus, saliva and breast milk.

macrophage – A type of white blood cell.

malleable – Can be hammered or pressed into shape.

mammogram – An X-ray image of the breast tissue.

mean – The average value, calculated by adding up all the numbers, then dividing by how many numbers there are.

measuring wheel – A hand-held device for measuring longer distances.

medium – A material, such as air, water or a solid, that sound travels through.

memory cells – White blood cells left in the blood after infection. They make antibodies quickly if you are infected again by the same microorganism.

menopause – The point at which a woman stops being able to reproduce, as her ovaries stop releasing eggs.

metabolism – All of the chemical changes in the body, especially those used to provide energy.

microorganism – An organism that can only be seen with the aid of a microscope.

mirror – An object that reflects light in a way that produces a virtual image.

monomers – The small molecules that join together to make polymers.

motion – Movement that results in an object changing its position.

mucus – A sticky fluid produced by your body to trap particles.

multimeter – An instrument that can be used to measure voltage, current, resistance and sometimes frequency.

nanochemical – A substance made up of nanoparticles.

nanoparticle – A particle that has one dimension which is less than 100 nanometres (nm).

nanotube – A tiny tubular structure formed by a giant lattice of linked carbon rings.

nanowire – Single crystals of carbon that conduct electricity such as graphite, but are very small, thin and light.

natural acquired immunity – Immunity acquired by being infected by a pathogen and overcoming that infection.

negative correlation – Data show a negative correlation if the dependent variable decreases as the independent variable increases.

negative temperature coefficient (NTC) thermistor – An electrical component whose resistance decreases as it gets warmer.

neutralisation – A reaction in which an acid reacts with a base to form a neutral salt and water.

newton (N) – The unit of force.

newton meter – A device used to measure force.

non-specific immune response – Immune response which is non pathogen-specific; i.e. involving physical barriers, chemical defences, inflammation and phagocytosis.

non-steroidal anti-inflammatory drugs (NSAIDs) – Drugs used to reduce inflammation.

non-uniform motion – An object moving with velocity or acceleration that is changing.

normal force – A force acting at 90° to a surface.

normal line – Imaginary line which is drawn at 90° to the surface of a mirror.

nutrients – Substances required by an organism for growth or energy.

obesity – A medical condition in which someone is very overweight, due to an abnormal increase in the number of fat cells around internal organs or under the skin, and has a BMI over 30.

Ohm's law – States that, at constant temperature, the current through a conductor is directly proportional to the voltage across it. The relationship is represented by the equation: voltage = current × resistance ($V = I \times R$).

opioids – Drugs made from the opium poppy, or synthetic chemicals having similar properties to these drugs.

optic nerve – A nerve that carries information from the eye to the brain.

optical fibre – A flexible transparent fibre that transmits light using total internal reflection.

organ donation – Giving of an organ by a donor for transplant into a recipient. People can opt to donate organs after they die or can give some organs, such as one kidney or part of the liver, while they are living.

organic molecule – A molecule containing carbon atoms obtained from a source that was once alive, for example crude oil.

oxidation – A reaction in which a molecule gains oxygen atoms.

palliative care – The care given to terminally ill patients; it involves pain relief and support to patients and their families.

parallel circuit – A circuit in which electrical components are connected so that the same voltage is applied to each component.

pathogen – A microorganism that can invade other cells or organisms and cause disease.

periscope – An instrument for observation from a concealed position. Its simplest form consists of an outer case with mirrors at each end set parallel to each other at a 45° angle.

phagocytes – White blood cells that protect the body by ingesting harmful foreign particles, bacteria, and dead or dying cells.

phagocytosis – The process carried out by phagocytes when they ingest a harmful particle, such as smoke in the lungs, or a pathogen, such as a bacterium or virus.

phenylalanine – An amino acid used to make melanin; causes severe learning difficulties if it builds up in the body.

phenylketonuria (PKU) – A rare inherited disorder where sufferers cannot process phenylalanine; causes irreversible brain damage if not treated.

pie chart – A chart in the form of a circle in which the size of the sectors is proportional to the frequency of the data.

plagiarism – Using other people's work and presenting it as your own.

plane mirror – A mirror with a flat surface.

pluripotent – Able to differentiate into many cell types.

polymer – A long molecule formed when many small molecules, called monomers, bond together.

polymerisation – The reaction in which a polymer is formed.

positive correlation – Data show a positive correlation if the dependent variable increases as the independent variable increases.

precision – (of a measuring instrument) The smallest change which an instrument can measure.

pre-eclampsia – A condition which can occur during pregnancy. Symptoms in the mother include high blood pressure, fluid retention and excess protein in the urine. It can cause growth problems in the unborn child.

principal axis – The imaginary horizontal line through the centre of a lens from which the size of objects and their images are measured.

prism – A transparent optical tool with flat, polished surfaces that refract light.

products – The substances produced by a chemical reaction.

prognosis – Prediction of the likely outcome of an illness.

prostate gland – An organ in the male, situated near to the urethra. It produces fluid that bathes sperm.

qualitative – Describes things that can be observed, without using numbers.

quantitative – Describes something using numbers.

rarefaction – Area where molecules are spread out. This will be an area of low pressure.

ray diagram – A diagram that shows the paths of rays of light as they interact with an optical instrument such as a lens or mirror.

reactants – The substances that react together in a chemical reaction.

real image – An image that can be projected on a screen.

recipient – Person receiving; in this context, receiving a blood transfusion or a body organ.

reflected – Rays or waves are reflected when they bounce off a surface, for example light bouncing off a mirror.

reflecting telescope – A telescope in which the focusing of the main image is done by a curved mirror.

refracting telescope – A telescope consisting of a series of lenses.

refraction – The bending effect that occurs when light enters or leaves transparent materials.

regenerative medicine – Using stem cells to replace or regenerate human cells, tissues or organs to restore normal function. This could solve the problem of too few donors for patients needing organ transplants. It also reduces the risk of rejection.

rejection – Immune response of the recipient that causes deterioration and death of the transplanted organ.

repeatable – Results are repeatable if, on repeating the investigation, you get the same or similar results.

reproducible – Results are reproducible if, when you change the method or use different equipment, or if someone else does the investigation, the results are still similar.

resistant – Bacteria that become resistant to antibiotics have the ability to survive treatment with antibiotics.

resistor – An electrical component that limits current flow.

resultant force – The net (overall) force acting on an object.

retina – Tissue at the back of the eye that contains light receptors.

Rhesus factor – An antigen found on the surface of red blood cells.

risk – The harm that could be caused and the chances of it happening.

series circuit – A circuit in which electrical components are connected so that the same current flows through each component.

sickle cell disease – A genetic disorder caused by inheriting two copies of the recessive allele of the gene for making haemoglobin. It causes tiredness, shortness of breath and periods of extreme pain in the joints.

significant figures – The digits in a number which give information about the precise value of that number.

smart material – A substance that has properties which change in response to an external stimulus, e.g. light or heat.

soluble – Able to dissolve in a particular solvent, usually water.

solute – A dissolved substance.

solution – A liquid formed by dissolving a solute in a solvent.

solvent – A liquid in which other substances (solutes) will dissolve to make a solution.

sonar – A way of finding the distance to an underwater object (such as the sea floor) by timing how long it takes for a pulse of ultrasound to be reflected from the object.

specific heat capacity – The energy (in joules) needed to heat up 1 g of a substance by 1 K.

specific immune response – Involves identification of specific antigens on a pathogen, and production of antibodies, to prevent the pathogen entering your body.

static frictional force – The frictional force acting on a non-moving object.

stopping distance – Thinking distance + braking distance.

structural formula – Shows all the atoms in a molecule, but does not show the bonds.

suspension – Undissolved particles in a liquid.

tensile force – A pulling or stretching force.

terminal velocity – A constant, maximum velocity reached by objects falling. This happens when the weight downwards is equal to the air resistance upwards.

thermal energy – Energy transferred by heating.

thermistor – A component whose resistance varies with temperature.

thermite reaction – The highly exothermic reaction of aluminium with iron oxide to produce iron and aluminium oxide.

thinking distance – The distance travelled by a vehicle while the driver reacts to a hazard.

tidal volume – The volume of air breathed in and out in each breath.

total internal reflection – Occurs when light passes from one material to another and the angle of incidence reaches a critical size, so that refraction is not possible and all the light is reflected.

totipotent – Able to differentiate into all cell types.

toxic – Poisonous.

transplant – Replacement of a diseased organ in the body of a recipient with a healthy organ from a donor.

ultrasound – Sound that is too high for humans to hear.

unbalanced forces – When forces are unbalanced, there is a resultant force and the object will change speed.

unethical – Morally wrong.

uniform motion – An object moving with velocity or acceleration that is not changing.

universal donor – A universal donor can donate to anyone; for example with blood donation, people with the blood group O can donate to everyone.

universal recipient – A universal recipient can receive from anyone; for example with blood donation, people with the blood group AB can receive blood from everyone.

upright image – An image that is the right way up.

vaccination – Introduction of a vaccine into your body to make it carry out a specific immune response and protect you from the disease the pathogen causes.

vaccine – A mixture containing a weakened or killed form of a disease-causing microorganism, which is injected into people to make them immune to that disease.

vascular – Related to the blood vessels.

velocity – Speed in a particular direction.

virtual image – An image that cannot be projected on a screen, for example when a convex lens is used as a magnifying glass.

viscous – Thick, slow-flowing; the opposite of runny.

voltmeter – An electrical instrument that measures voltage.

wavelength – The distance between a point on one wave and the same point on the next wave.

work done – A type of mechanical energy, explained by the equation: work done = force × distance moved in the direction of the force.

zygote – A fertilised egg cell.

Index

The publisher would like to thank the following for their kind permission to reproduce their photographs:

(Key: b-bottom; c-centre; l-left; r-right; t-top)

Alamy Images: Corbis Cusp 120t, culture-images GmbH 68, fStop 75t, Golden Pixels LLC 41, Health and Medicine 156, Huntstock Inc 155, Ingram Publishing 146, Jinx Photography Brands 39r, Martin Jenkinson 59r, Momentum Creative Group 84, Moodboard 153, Peter Alvey 38, PhotoAlto sas 97, Science Photo Library 176, Sandy Young 80; **Ariane Space:** 51; **British Liver Trust:** 113l; **Corbis:** Adam Gault / Science Photo Library 85, David Harrison / Able Images 54, Gideon Mendel for The International HIV / AIDS Alliance 161, HBSS 25, Hybrid Images / Cultura 29, John Sommers / Reuters 80r, Ralph Hutchings / Visuals Unlimited 149, Science Photo Library 20t; **Courtesy of CitizenCard:** 112; **Fotolia.com:** Auremar 99, Danny Hooks 23b, David Asch 59l, Goodluz 67, Jessmine 72, Maksym Dykha 55b, Nikesdoroff 34, Plus69free 151, Ruslan Solntsev 35t, Russell Witherington 59c, Secret Side 78, Snow White Images 39l, Suprijono Suharjoto 143; **Getty Images:** 75b, Henrik Sorensen / Riser 105, Martin Bureau 174, Roger Harris / Science Photo Library 117l; **Imagestate Media:** John Foxx Collection 61; **NASA:** 17, SAO / CXC 81r; **NHS Blood and Transplant:** 139; **Pearson Education Ltd:** Studio 8 93, 104, Gareth Boden 96, Trevor Clifford 19t, MindStudio 152, HL Studios 20b; **PhotoDisc:** Jack Star / Photolink 133, Photolink 69, Ryan McVay 119; **Rex Features:** Jennifer Jaquemart 60; **Science Photo Library Ltd:** A J Photo 137, 116r, 130, 173, Andrew Lambert 31l, 31c, 31r, Andrew Lambert Photography 32, 49, 79tl, 79tr, BSIP VEM 126l, BSIP, Bajande 122, BSIP, Laurent 126r, Carol & Mike Werner / Visuals Unlimited, Inc. 117r, CNRI 113tr, 113br, Dr Fred Hossler, Visuals Unlimited 115, Eye of Science 116l, GE Medical Systems 120, General Electric Research and Development / Emilio Segre Visual Archives / American Institute of Physics 178b, Geoff Kidd 132t, J.L. Martra, Publiphoto Diffusion 138, John Durham 128, Kings College School of Medicine 124l, Edward Kinsman 87, Look At Sciences 120b, Martyn Chillmaid 55t, 150t, Mischa Keijser 77, Cordelia Molloy 79b, NASA / Chris Gunn 81l, NIBSC 114, Paul Gunning 129, Philippe Plailly 159, PR.M. Brauner 111, RIA Novosti 148r, Samuel Ashfield 124r, Saturn Stills 118, 177, Sheila Terry 95, Simon Fraser / RVI, Newcastle-Upon-Tyne 123t, St Mary's Hospital Medical School 107, Steve Gschmeissner 140, Victor Habbick Visions 42, 141, 147l, Charles D Winters 14, Peidong Yanc 11, Zephyr 142; **Shutterstock.com:** Anatoliy Samara 108, Carlos Wunderlin 88, Parpalea Catalin 35b, Fedorov Oleksiy 150b, forbis 132b, Dana Heinemann 47, Inga Ivanova 36, 45, Jabirn 160, JJ Studio 76, Jurand 24, kRie 71, Li Wa 135, Lightpoet 145, Mariano N. Ruiz 40, Michal Kowalski 158, Mikael Damkier 162, Nigel Paul Monckton 23t, Pedro Salaverría 154, Sashkin 178t, Scott Rothstein 96t, Sergey Furtaev 144, Sianc 50, Supri Suharjoto 10, Tramper 57; **Sozaijiten:** Sozaijiten 148l; **SuperStock:** Simon Wilson 86; **TopFoto:** Peter Huizdak / The Image Works 58; **Veer/Corbis:** Keng Po Leung 73l, Media Image Photography 147, Monkey Business 19b, 123b, Monkey Business Images 73r, okea 30, onepony 98, plampy 89, Sarah Allison 64

Cover images: *Front:* **Getty Images:** Adam Gault / OJO Images

All other images © Pearson Education

In some instances we have been unable to trace the owners of copyright material, and we would appreciate any information that would enable us to do so.